INTERFACING OURSELVES

Interfacing Ourselves consists of new work that examines digital life on three levels: individuals and digital identity; relationships routinely intertwining digital and physical connections; and broader institutional and societal realities that define the context of living in the digital age. A key focus is what it means in varied social arenas when most individuals live as *co-present* or *multi-present*—simultaneously engaged in digital and physical space—alone and with others. Topics include how: digital life contributes to well-being; individuals experience digital dependency; a smartphone is more than a smartphone; netiquette reveals social change; some online communities become prosocial salient havens while others reinforce social inequality; Millennials build intimacy; Latinx do *familismo*; and digital surveillance and big data redefine consumerism, advocacy, and civic engagement. Six chapters incorporate insights from hourly journals of Millennials undergoing a period of digital abstinence. Other chapters draw from surveys, digital auto-ethnography, content analysis, and other methods to explore digital life at the level of individual and interactive experience, and at a broader institutional and societal level. Ultimately, the book presents the need for living a mindful digital life by developing greater awareness as an individual, a social being, and a netizen and citizen.

Cristina Bodinger-deUriarte is Professor of Sociology at California State University, Los Angeles (CSULA), where she was named "Woman of Distinction." She teaches popular culture, social psychology, and research methods, while serving as MA Program Director, Associate Chair, and Webmaster. Her background includes a Harvard PhD, educational program evaluation, government contract research, and business consultation. Her book, *Hate Crime*, won a Rising Star award.

INTERFACING OURSELVES

Living in the Digital Age

Edited by
Cristina Bodinger-deUriarte

Routledge
Taylor & Francis Group
NEW YORK AND LONDON

First published 2019
by Routledge
52 Vanderbilt Avenue, New York, NY 10017

and by Routledge
2 Park Square, Milton Park, Abingdon, Oxon OX14 4RN

Routledge is an imprint of the Taylor & Francis Group, an informa business

Library of Congress Cataloging-in-Publication Data
A catalog record for this title has been requested

ISBN: 978-0-367-23506-2 (hbk)
ISBN: 978-0-367-23510-9 (pbk)
ISBN: 978-0-429-28010-8 (ebk)

Typeset in Bembo
by Taylor & Francis Books

TO GUNNAR, WHO IS MY PARTNER IN EVERY WAY, FOR HIS UNWAVERING KINDNESS, ENCOURAGEMENT, AND COLLEGIALITY.
ALSO TO MY CURRENT AND FORMER GRADUATE STUDENT COLLEAGUES
—EACH A UNIQUE SOURCE OF INSPIRATION—
FOR HELPING KEEP MY MIND ACTIVE AND FLEXIBLE.

CONTENTS

TABLES

EDITOR'S NOTE ON THE CONTRIBUTORS

I developed an ongoing colleagueship with the contributing authors when we crossed paths in the CSULA Sociology Department. Three of us are currently CSULA faculty members. Before moving on, however, the other six had each won the *Service to the Department Award* based on their instructional contributions. Even after leaving, three of them (Berge Apardian, Grzegorz Hryniszak, and Daniel Okamura) continued working with Gunnar Valgeirsson and me through the completion of the digital deprivation study—a study contributing to several of these chapters. Deborah Burns also worked on the earlier wave of this study with us. Collectively, we have earned degrees in sociology, psychology and communications, and have taught (among other topics) classical and contemporary sociology theory, popular culture, research methods, content analysis, social problems, social psychology, and the sociology of media.

All of us have presented our work on digital sociology at conferences, three of us as keynote speakers. Most recently our digital sociology conference venues included:

- American Association of Behavioral & Social Sciences (AABSS)
- California Sociological Association (CSA)
- Mini-Conference: Digital Sociology (ESS affiliated)
- Eastern Sociological Society (ESS)
- Far West Popular Culture Association (FWPCA)
- Our Millennial Era: Living in the Digital Age (OMELDA)
- Pacific Sociological Association (PSA)

Berge Apardian, Washington DC Public Schools.
Cristina Bodinger-deUriarte, CSULA.
Deborah Louise Burns, Iowa State.
Wai Kit Choi, CSULA.
Grzegorz Hryniszak, Netflix (cultural assessment and translation)
Michael Nitzani, Pierce Community College.
Daniel Okamura, UNLV.

Sasha Santhoff, HIV Sexuality Study.
Lacey M. Sloan, Director of Social Services for LifeSTEPS Inc.
Gunnar Valgeirsson, CSULA.

INTRODUCTION

This is an Invitation

Cristina Bodinger-deUriarte

This book uses a collection of new, original works to examine what it means to "interface ourselves" by digitally mediating our identities, relationships, social expectations, consumption, activism, and politics. Much of the earlier scholarship on digital life has focused on the idea of virtual activity (*digital presence*) versus activity "in real life" (*physical presence*). The overall approach of this book agrees with Miller that "the practice of contrasting the virtual with the real is problematic, as 'reality' is made up of both the concrete and the virtual in many ways" (Miller 2011:33). The narrative across the chapters in this book treats our digital presence (online) as more significant than something that simply opposes our physical presence in *directly-inhabited* (offline) environments.

Extending beyond the idea of "real" versus "virtual," this book takes a *co-presence* perspective—acknowledging that physical presence and digital presence continuously intertwine, mediate, and influence one another. Chapters address questions such as:

- To what degree is our sense of a meaningful life, and how we fit into our environment, tied to digital as well as physical presence?
- When is a smartphone more than a smartphone?
- When is online community more salient than physical community?
- How do digital images and subcultures interact to shift subcultural expression?
- How do Millennials maintain love relationships and engage in intimacy-building?
- How do netiquette concerns reveal broader social change?
- When is online interaction not enough?
- How do millennial Latinx do *familismo* through presence simultaneity?
- What does it mean when someone claims to be an Internet addict?
- How are Millennials "winning" the music piracy debate?
- Do political websites dissipate activism?

- Should you worry about digital surveillance and data mining?
- How accurately does cyberspace reveal you to others?
- Just how much is too much, and how do we know?

In scholarship on digital life, three seemingly contradictory uses of the terms "co-presence" and "co-present" can lead to confusion. Consequently, the authors in this book share the same definition of co-presence:

> The practice of people engaging in real-time face-to-face social activities and being engrossed in the here and now of that social activity, but also being virtually present … with dual occupation of real and virtual space (IGI Global nd).

The second definition is: "being present and connected with others in a virtual environment" (Op. Cit.) In order to avoid confusion, this meaning will always be specified as "digitally co-present" or will be replaced by another phrase, such as "sharing digital space." The third definition is: "participants in a communication being located in the same physical setting" (Op. Cit.). Again, this will always be clarified by substituting a phrase such as "sharing directly inhabited space" or "physically co-present." When the terms co-presence or co-present are used in this book, they will always refer back to the first of these three definitions—that of being simultaneously sharing physical space while being engaged in a digital environment *not* being shared with the physical space companion(s).

This book is intended to function as a springboard for deep discussion and reflection around how we interface ourselves in our simultaneously-digital and non-digital lives. Consequently, authors take a reader-friendly, investigative and reflective tone while presenting insights that are theoretically grounded and empirically sound. In accord with this approach, theoretical concepts are unpacked as they appear within the chapters. Additionally, only passing reference is made to research methods *within the chapters*, leaving a more instructive set of critical reflections on the varied methodological approaches for the final book chapter. The variety of sociological perspectives and research methods give this book an authentic voice.

In order to reflect varied aspects of digital life, authors have drawn on the use of millennial journals, virtual observation of online communities, visual sociology, survey data, content analysis of website narratives and advice columns, and sociological phenomenology. Several chapters include a brief socio-historical background to help give the topic a meaningful context. A subset of six chapters addressing disparate topics include material from hourly journal entries made by 420 Millennials who were deprived of access to smartphones, computers, and other digital devices for twelve consecutive waking hours (in a time period of their own choosing). These 5,040 journal entries reveal myriad aspects of the essential and internalized nature of digital life.

Almost half of the chapters in this book focus specifically on the millennial generation; some additional chapters also include mention of millennial digital life. Millennials have been targeted, as the first generation of *digital natives*, those who were born into a world of personal computers and the Internet. They have also been the most digitally engaged for the greatest amounts of time, and use the most digital platforms. They tend to be more frequently *co-present*, inhabiting digital and physical space simultaneously. Given their centrality, it is important to define this millennial generation. This book uses the same perimeters as the Pew Research Center—an institute well-known for researching Millennials and digital behaviors for over a decade:

> In order to keep the Millennial generation analytically meaningful, and to begin looking at what might be unique about the next cohort, Pew Research Center will use 1996 as the last birth year for Millennials for our future work. Anyone born between 1981 and 1996 (ages 22 to 37 in 2018) will be considered a Millennial, and anyone born from 1997 onward will be part of a new generation (Dimock 2018).

The six chapters using Millennials' journals as data are drawing from hourly journal entries written across a set of years by college students, so the population was more concentrated among Millennials who were between the ages of eighteen and their middle-to-late twenties. None were over thirty-three.

Primarily, this book is an invitation to explore what it means to live in a world where digital presence and physical presence continuously interact. Readers are invited to consider all levels of digital life—from micro-social questions like "How many texts are too many?" to macro-social concerns like "How has digital surveillance and *big data* redefined the world in which we are consumers, advocates, and citizens?" Ultimately, the hope is to inspire readers to develop greater self-awareness in relationship to their co-presence as individuals, as social beings, and as world citizens and *netizens*.

To facilitate focus on the topical content, rather than the process, research methodologies were almost completely omitted from the topic-focused chapters. Instead, they have been collected in the fifteenth chapter. The methodology chapter includes process reflections and annotated highlights for those who wish to more deeply consider the realities of data collection.

Theoretical Orientations

This book unfolds within a social systems approach. Digital life is considered in terms of both *micro-social* and *macro-social* contexts.

Chapters addressing *micro-social* topics discuss the experiences of everyday life. Micro-social contexts are generally those we interact with most directly. They may have the greatest impact on our own development as individuals and include family, friendship circles, romantic relationships, work sites, interest-based groups, and other direct-involvement settings.

Social media and virtual communities expand the notion of micro-social contexts by including digital settings as "direct-involvement settings," even though interactivity takes place in a non-delimited space. Digital space blurs the boundaries *among* micro-social settings, and the members of those settings. For example, friends and relatives cross virtual paths far more than they would have interacted in physical settings. From a micro-social approach, individuals are treated as *social actors*, whose perspectives, expectations, and the meanings they attribute to interactions with other individuals and groups, have important social implications. Within this context, digital presence is approached as interpenetrating and inseparable from physical presence. Concerns center on social expectations, norms, degrees of self-awareness, the sense of well-being, and other aspects of life as it is lived. Ultimately, the concern is how individual choices and decision-making around digital and face-to-face behavior have implications that individual social actors may not fully recognize. Chapters 1 through 5 share a focus on the individual and individual development. The following four chapters, 6 through 9, focus on interactions and relationships among people at the micro-social level, while still considering broader social implications.

Chapters 10 through 13 address *macro-social* topics. These four chapters focus on broader *social forces* associated with the major social structures and institutions that shape the society in which we all live. In this context, the primary question can loosely be framed in this way: How do the realities of digital space environments reflect the interests, control strategies, and power relations of social institutions in view of countervailing attempts to protect or elevate individual *social agency*? Political, economic, and enforcement institutions, for example, may exert power in ways that have far-reaching effects on individuals navigating digital space. Increasing issues are raised around the control and use of digitally-provided information. For example, how the third-party use of Internet data may undermine true social agency—that is, the degree to which a person can act independently and exert free will. Concerns still involve social expectations, norms, degrees of self-awareness, and individual choices and decisions. However, this is addressed in consideration of major social institutions making choices and decisions that affect individuals in ways that individuals may not fully recognize.

The Internet is unique in how it manifests both micro-social and macro-social levels in a dynamic way. It provides avenues for social control, surveillance, enforcement, and the commodification of human interaction, while simultaneously providing the potential for social change, civic engagement, resistance, and consciousness-raising.

Multiple theoretical and conceptual lenses are used to examine different topics. Chapters exploring micro-social phenomena tend to take a social psychological approach, drawing on symbolic interaction concepts. On the other hand, conflict theory dominates the approach in chapters exploring broader macro-social forces. It is important to note, however, that these primary theoretical orientations do not act to contradict one another, but dovetail. Although sharing some common ties to social psychology and/or conflict theory, chapters are distinguished from one another by the particular set of conceptual lenses through which they view the subjects they discuss. These include aspects of social integration, feminist theory, and queer theory, as well as applications from the continuum of perspectives whose endpoints are the social construction of technology and technological determinism. Although theoretical concepts ground each chapter, the focus of the book is not heavily theoretical. Rather, theoretical underpinnings are used to illuminate. They are explained and made accessible in the Part I–IV prefaces and in individual chapters in order to keep primary attention on the actual phenomena as experienced in everyday life. Further, although the focus and theoretical lens may vary, chapters tend to share an emphasis on: (1) informed, self-aware decision-making around individual digital presence; and (2) understanding the broader implications of living in a digital era.

As a collection, these chapters address both micro-social and macro-social issues. The balance and interpenetration between physical presence and digital presence is a critical feature of both micro- and macro- social contexts. Chapters are grouped by micro- and macro-social orientation. However, online the micro-social and the macro-social can combine to reinforce one another, or can come into conflict, so some of these boundaries overlap. Taken together, the chapters contribute to an understanding of how the internalized norms at the micro-social level, reflect and help construct new societal realities at the macro-level. Similarly, the chapters illustrate how powerful interests in the formation, later development, and socio-political and economic use of the Internet, constitute social forces that influence and control aspects of human experience and affect individual and collective social agency.

Part Focus

The book is divided into four parts, each of which is preceded by a brief part preface that describes the overarching orientation of the part, any shared thematic strands, and the theoretical underpinnings (whether elaborated or briefly referenced) across that section of chapters. Part prefaces also explain some of the specialized terms in the upcoming chapters. The commentary below is, therefore, more of a quick overview to finish explaining how this book works as a whole.

Part I, "Digitizing Identity," has a micro-social orientation that looks at effects on, and significance for, *the individual*. The emphasis is on how individuals "do" co-presence—particularly through social media and group forums—in view of individual well-being, identity and socialization, the competition of values, and social integration.

Part II, "Mediated Relationships," has a micro-social orientation that focuses on the *interaction between or among individuals*. Topics include romantic relationships, family dynamics, and friendship. These chapters consider individual choices about the balance between physical and digital presence in view of relationship implications and social repercussions.

Part III, "Virtual Agency and Digital Dystopia," has a macro-social orientation that focuses on the struggle between *social forces* and *social agency*. Topical areas include: social change; civic commodification; enforcement; data mining and filter bubbles; civic engagement; digital surveillance, and social justice. These foci are united by their treatment of power struggles; the control of informational flow; the influence on personal perceptions given the nature of accessible information about users; and the invisible/semi-visible use, manipulation, and application of digitized data.

The goal across chapters in these three parts is a thought-provoking gestalt that underscores a need for individuals to bring conscious reflection to their digitally-mediated choices—one in which they understand the visibility of their online presence and its impact. Together, these chapters demonstrate the importance of continuing to examine the broader implications of how we interface ourselves as physically-present and digitally-present social actors, what it is to be *mindfully* co-present, and how we locate ourselves in the struggle between social control and social agency.

Part IV, "Capping it Off," provides two culminating chapters that are reflective and pragmatic. Chapter 14 takes the form of a reflective capstone, but introduces further insights and information. Millennials who participated in the twelve-hour digital abstinence experience provided additional reflections not previously addressed in the book. These are presented in view of the key questions they pose, the insights they offer, and the important considerations they raise. Chapter 15 has an applied focus. It provides an annotated review of the methods used across the first thirteen chapters. Author methodologies are placed in a broader context in view of the overall methodological orientations and approaches in general, not only in terms of how methods were specifically applied by particular authors. Relative merits and particular challenges are addressed.

References

Dimock, Michael. 2018. "Defining Generations: Where Millennials End and Post-Millennials Begin." *Pew Research Center*March 1. Retrieved June 26, 2018 (http://www.pewresearch.org/fact-tank/2018/03/01/defining-generations-where-millennials-end-and-post-millennials-begin/).

IGI Global. nd. "What is Co-Presence?" Retrieved September 10, 2018 (https://www.igi-global.com/dictionary/co-presence/4115).

Miller, Vincent. 2011. *Understanding Digital Culture*. Los Angeles, CA: Sage.

Part I

DIGITIZING IDENTITY

Preface

These five chapters examine social actors in micro-social context. A common thematic strand is how online (*digital presence*) and offline (*physical presence*) combine as co-presence to provide the framework for self-development, self-awareness, and individual choices. This part also considers how the *collective consciousness* of Millennials—that is, their identification as a cohort with particular shared understandings and expectations—contributes to their sense of *social integration*, or their sense of belonging in the world. Given the importance of being co-present in physical and digital space, Chapters 1 and 5 explore what we can learn from the reactions of Millennials in a state of *digital deprivation*. Chapter 1 considers the importance of meaningful co-presence while Chapter 5 looks at what it means when a subset of digitally-deprived Millennials self-identify as digitally dependent or "addicted." Both chapters draw from the hourly journals of 420 Millennials who remained offline for a period of twelve consecutive waking hours (at a start time and day of their choosing).

Contributing theoretical outlooks are often implicit, rather than specifically articulated. In such instances, the material focus is emphasized. Nonetheless, all chapters have strong foundations in theory. Part I's chapters primarily draw insights from social psychology. Symbolic interaction concepts and social constructionism are particularly well-reflected. Concepts are explained at a basic level within each chapter, as needed.

Chapter 1 uses both psycho-sociological and classical sociology theory concepts. The former includes social psychology approaches to basic human needs for psychological well-being; perceptions of a meaningful life; and the development of coping strategies during times of psycho-social deprivation (including digital deprivation). The latter includes concepts of collective consciousness and social integration. Consequently, those wishing to explore the ideas in more depth can investigate this variety of concepts beyond the fundamental descriptions provided in the chapter.

Chapter 2 is built upon theoretical underpinnings combining, the role of media in reinforcing mainstream and subcultural values, socialization, social constructionism, social agency, and feminist theory. The narrative is more focused on the material, or concrete phenomena, however. Readers interested in a deeper appreciation of these theoretical foundations may consider how reference groups, and the social agency of the tattoo artists combined across different eras to change public opinion and social norms related to tattoos. Another concept of interest for this chapter is that of *moral panic*—the mass rejection of a social behavior due to its perceived link to moral decay or deviance. The role of media as an agent of socialization is also key, as are feminist insights about the nature of gender inequality in the online tattoo culture. These combine to illuminate issues such as the internalization of gendered expectations, male-dominated tattoo cultural arenas, the objectification of females, self-objectification, and the resistance to objectification in digital-space tattoo communities.

Chapter 3 stands out in this part as specifically theoretical in its focus. It presents different theoretical approaches from which to consider the relationship between technology and the users of technology. The key issue is the degree to which the individual has *social agency*—meaningful power over decision-making and behavioral choices—in view of the technological features and expectations related to smartphones. Each theoretical frame is associated with a different level of acknowledged social agency. The four are: technological determinism (TD), social construction of technology (SCOT), actor-network theory, and the social shaping of technology (SST). Although primarily focused on theory, the explanations are quite accessible and are illustrated with commentary from the millennial journals.

Chapter 4 has a strong mental health focus in a queer theory context. Broader consideration of *queer theory* involves understanding notions of *heteronormativity* as embedded with power and control that applies pressure to both straight and LGBTQ+ individuals. The social psychology theoretical foundation draws strongly on Goffman and on the sociology of emotions. Those wishing to explore the sociology of emotions concepts informing this chapter might look at literature on emotion management, coping mechanisms, and intimacy-building through self-disclosure. The author approaches these concepts in view of normative pressure in physical space (as Goffman's front stage), and social expectations in specialized digital-space communities for LGBTQ+ individuals (as Goffman's backstage). This involves the concept of virtual community, and the social psychological concepts of reference group, self-presentation, image management, and masking and unmasking—all of which are clarified in the context of the chapter.

Chapter 5 has a more *phenomenological* approach; that is, it has more of a focus on life as experienced by the participants. Those interested in this more socio-philosophical approach would need to consider how individual consciousness, expressed in the first-person point of view, indicates self-awareness in relationship to digital life. There is significant overlap with the social construction of meaning, as seen in symbolic interactionism. In this case, the key issue is how digital technology users self-define as addicted to the Internet. The author draws on established behavioral indicators in the psychology of addiction to compare these phenomenological, constructivist self-definitions with mental health perceptions on behaviors that reflect addiction.

1

OFFLINE AS MISALIGNED

Millennials Coping with the Loss of Digital Presence

Cristina Bodinger-deUriarte

Relative Deprivation and Internet Intensity

A recent survey reports that a significant proportion of post-Millennials (teens) believe they could give up their phones for at least one full day (eMarketer 2017). Current study results cast doubt on this claim. All 420 millennial college student participants (primarily in their twenties) said that they could give up their phones and all Internet access for twelve consecutive waking hours, yet over half (60%) wrote about intensely craving their phones and about one-quarter (24%) reported giving in and checking their phones one or more times. Participants had very strong reactions to being unplugged for a half-day period (despite being able to choose the start time and, within limits, the date). This chapter draws on their 5,040 hourly journal entries to discuss why they were so deeply affected by the experience. To truly understand their responses to digital deprivation requires one to recognize that being online means more than simply engaging in a set of activities. One must also consider more complex explanations than the popular narrative of wholesale addiction or generational narcissism. Before discussing the meaning of digital involvement for this population, however, it is important to consider the intensity of Millennials' digital presence.

Internet Intensity

For US Millennials, the highest intensity of Internet usage involves social media. Smith and Anderson (2018) identified the eight most popular digital platforms: YouTube (94%), Facebook (80%), Snapchat (78%), Instagram (71%), Twitter (41%), Pinterest (34%), LinkedIn (29%), and WhatsApp (27%). The ninth and tenth would be Google+ (26%) and Tumblr (21%) in data from Statista (2018). Further, Smith and Anderson (2018) found that

Millennials under thirty use four of these platforms regularly—generally multiple times a day—as is true for 71 percent of Snapchat users and 55 percent of Instagram users. Almost all Millennials (92%) own a smartphone (Jiang 2018) and three-quarters of them check their phones between 50 and 100 times daily (Globe Newswire 2017). Thirty-nine percent of survey respondents in their late teens and twenties report being online "almost constantly" (Perrin and Jiang 2018). One-quarter of US Millennials spend about one-third of their waking hours on the phone (Globe Newswire 2017). This intensity helps explain why 85 percent of smartphone owners always keep their smartphone in direct reach, and most take it to bed with them.

Co-Presence as More Than an Either/Or Proposition

Digital activity frequently takes place while being physically present with friends and family. Americans across age groups are increasingly engaged in this kind of co-presence; the majority (85%) report using the phone while also talking with family and friends who were physically present (Richter 2018).

Millennials are often portrayed as shifting distractedly from online entertainments and contacts to offline interactions, dividing their attention between competing interests in a "virtual world" of simulated and mediated interactions, and a "real world" of authentic relationships. The strength of their ties to digital space platforms and interactions has often been depicted as existing at the cost of their presence in directly inhabited, physical space environments. The author argues, however, that the social-psychological issues are complex, and that important insights are available with a different approach than this "either/or" treatment of digital presence versus physical presence. A more holistic approach supports the observations of a number of digital research scholars (Miller 2011; Turkle 2011); however, the co-presence/multi-presence approach taken here is still underrepresented in sociological research.

What does it mean to sidestep the idea that digital space and physical space are experienced separately, like the rapid shifts of attention in multitasking? First, it is clear that digital devices extensively redefine and mediate experience in directly inhabited, physical environments. The following are just a few examples: apps that advance-order and pay for take-out coffees to avoid lines and interactions with baristas and other strangers; "Find My Friends" apps to meet up easily; apps to browse other "singles" in a shared physical location (like a bar) making approaching someone easier; apps to have the phone ring to "fake" getting a call in order to claim "something came up" so as to leave an undesirable interaction; and so much more.

Second, digital space environments are similarly created, shaped, and "managed," by incorporating narratives and comments, and by posting visual representations of events experienced in physical space. These narrative and visual posts, in turn, become the basis for further digital and physical interactivity (likes, comments, re-postings, related link sharing, meetings, etc.). Rather than balancing two dimensions of experience, Millennials exist multi-dimensionally. In fact, this research shows that online *digital presence* and offline *physical presence* interpenetrate to the degree that a simultaneous digital/physical *co-presence* or *multi-presence* is more than a lifestyle choice for most Millennials. This multi-presence has become an internalized aspect of selfhood and collective consciousness among Millennials.

The participants in this research kept journals about what digital activities they gave up (if any) each hour, what they did instead, and how they felt during each hour. After completing twelve hours of abstinence, some also appended reflections about the experience overall (see Chapter 14). The emotional and behavioral reactions, face-to-face interactions, and responses to their directly inhabited, physical environments, were impacted to an unanticipated degree. An analysis of reported moods and attempts to cope with digital deprivation reveals that suppressing digital presence is akin to suppressing key aspects of identity. It also creates a disconnect from collective consciousness and social integration. The end result is deprivation that seems to harm, at a basic level, this population's sense of well-being and of living a meaningful life.

Well-Being at Three Levels

Social psychology recognizes three particularly important social factors among the primary contributors to a feeling of well-being: (1) a general feeling of affiliation; (2) emotional intimacy; and (3) a sense of belonging (Erber and Erber 2001). Millennials use social media to support all three.

The Need for Affiliation

Affiliation is casual day-to-day interaction. This generally involves superficial relationships with acquaintances. In fact, many of one's social media friends fall into the category of affiliates, as opposed to close or intimate friends. Affiliation interactions focus on everyday activities, work, specific interests, or on simply moving through a typical day. Social media posts generally include this sort of material and journal entries included numerous examples.

> I was actually thinking of taking a picture of a drink which has no importance to me in this world.
>
> I wanted to text a colleague to gossip about something that was said at the meeting and knew that I would forget to do this later when I was allowed access to my phone.
>
> I am what most people would call a social media girl. I love social networking with other people, and I love updating my Facebook status or putting up new pictures of what I am doing at any time of the day.

Likes and comments received from social media friends or followers provide the casual sort of interaction associated with affiliation. However, simply posting the materials helps fulfill the need for affiliation because the person posting items imagines the reactions of others. This suits the normally superficial nature of affiliation. Researchers have found that college students who posted more status updates than they normally did felt less lonely over the course of a week, even if no one "Liked" or commented on their posts (Gannon 2013). This supports the idea that even without a response, affiliation activities contribute to one's sense of well-being and connection with the world.

The importance of digital space activity in satisfying this need is demonstrated when affiliating through social media is temporarily unavailable. Participants frequently mentioned their sense of "missing out" (31%), and being "isolated" (23%) even when others were sharing their physical space. Such feelings of displacement seem to indicate lost affiliation satisfaction:

> All I could think about was all the things I was missing online or if anybody had posted on my wall or texted me.
>
> By this time, I would have sent about three Snapchats that I can count (via my cell phone.) And possibly received a few, too. Instead I proceeded to cook breakfast without stopping to take a pointless picture of my scrambled eggs or my dog.
>
> I realized I had not been in such a bad mood in a long time. I can't remember the last time I felt so crappy, and isolated from the world. I thought staying away from technology was going to make me feel closer to people … but I was so wrong.
>
> I found myself becoming more and more isolated from the rest of society, mainly because I had no way of knowing what was going on in the world … My feelings of isolation and anxiety grew more and more because I felt as though I had been cut off from the rest of the universe—[was at home with her family].

Far from retaining confidence of their place in the world, one in six digitally-deprived participants described feeling "lost."

> This experience really affected my frame of mind. I completely felt out of place and I kept having the sensation of irritation and fear, due to me not being able to use my cell phone.

Need for Emotional Intimacy

Emotional intimacy satisfies a need to feel that we truly know another and that we are truly known by another. This entails the ability to share on a deep and meaningful level—it is typified by self-disclosure, honesty, and reciprocity; it requires trust and being trustworthy. Private messaging, in-jokes, symbolic memes, commemorative photos, customized avatar greetings, and other in-group postings are among the strategies people use to develop emotional intimacy in digital space environments. Still, it remains problematic to build intimacy in relatively-public venues. An unsatisfied need for emotional intimacy may, at times, motivate social media oversharing. Agger agrees, observing that "people overshare when communicating electronically in ways they wouldn't face-to-face. Embarrassment prevents them from saying things that they might text or tweet … Oversharing is not the same as intimacy" (2012:26). In other words, merely blurring the boundaries between public and private is not the same as developing reciprocated self-disclosure among trusted and trusting intimates.

Emotional intimacy was also seen as affected when digital connection through posts, texts and other messaging was curtailed. Participants reported distress over what they perceived as neglecting or hurting close friends and intimate partners.

> My boyfriend got sad … I text my boyfriend whenever I have time every day. I would have texted him on the way to the class but I could not. This made me feel sorry and guilty for my boyfriend and I got a little irritated.

Loneliness surfaced quickly despite the relationship being local and active.

> I gave up texting my boyfriend … I still felt lonely and imagined what he would be doing while I cannot contact him.

Despite having alerted close friends and intimates about their offline status, others often did not take this seriously. When some participants actually remained offline for any length of time, close friends and intimates expressed concern, some even suddenly visited to make sure participants were all right (even at their jobs).

The Need for Belonging

Belonging can be defined as feeling accepted and reasonably connected to social settings and to persons. This includes the idea that one's efforts, outlooks, and contributions are appreciated or have meaning. It includes the "one of us" feeling of identification. Reactions to social media profiles, posted observations, pictures, music, video clips, and other expressive material, enables one to build a sense of belonging through acknowledgement and inclusion. Participants frequently made such observations:

> It is millennial technology that is enabling people to have a sense of belonging and have some satisfaction about their existence as

> well as maintain relationships with others. It has become an important part of the communication tools that we cannot live without in order for the human species to remain a "social animal."
>
> Technology makes it easy for us to communicate to others and feel that we belong with them. That's why society created all these social networks and dating websites.
>
> The sense of belonging gives us reassurance that we are important and we are not outcasts. Social networks like Facebook, Twitter and dating websites help people find that connection to others by having access to their daily lives via online 24/7.

Losing one's sense of belonging typically leads to feelings of alienation or marginality. This happened very quickly and was reported in journals within the first few hours of digital abstinence.

> Not having technology in my life for those few hours made me feel like an outcast and as if I was detached from the world. While my friends were texting and using their phones I just felt like I didn't belong.
>
> I felt like I wasn't connected with my friends since I couldn't take any pictures or videos of myself while I worked out. Most of my friends do the same thing, which makes me feel like I belong in their group.
>
> Seeing other people use devices that I am accustomed to using myself, made me feel out of place. I wanted to feel like I belonged, but in a sense I didn't … Instead of belonging, I felt a sense of alienation.
>
> Usually I'm on my phone or have the headphones plugged in so that I don't seem to acknowledge them but [giving that up] I feel I was impacted greatly because it seemed as if I wasn't a part of my group.

Loss of Centrality

Interacting through customized social media platforms, gives individuals a specialized sense of belonging that includes being at the center of things. This happens because social media environments are customized and largely reactive to one's personal observations, events, and mundane daily moments. Participants frequently exhibited escalating anxiety as they felt a growing a sense of "peripheral" rather than "central" status among friends during the twelve-hour span.

> Whether it's checking in at restaurants on Facebook, posting pictures on Instagram, or just texting, we are on our phones. This

> time however, I was not. I engaged in conversation more and felt annoyed when my party would be on their phones.
>
> When I was taking the run with my friend, she was listening to her music while I was trying to talk to her, and she didn't really seem like she wanted to talk. It made me in a sense feel like I didn't belong, as opposed to "belonging," because she didn't seem appreciative of my contribution to our "mini" conversation. Ultimately this definitely made me feel out of place.
>
> All my friends were completely involved with their phones … I even told them about the game where everyone puts their phones stacked in the middle of the table and the first one to reach for their phone pays for everyone's dinner; they didn't get the hint.
>
> I kept talking to my friends regardless if they paid attention or not because I just don't want be quiet and not be part of the group … but maybe they got annoyed because I talked more this day just to get their attention since I didn't have my phone.

The disrupted sense of belonging was multi-dimensional. The inability to participate in digital life undermined the belongingness gained from digital activity and interactivity. Belongingness was also diminished by an awareness that their offline status was visible to others, and marked them as "not one of us." Finally, belongingness was diminished through lost feelings of connectedness and centrality with co-present friends in physical space.

Coping Strategies

Social media may support affiliation activities, provide venues for emotional intimacy, and enable a sense of belonging. The current research indicates, however, that for many, these are not anchored strongly enough to survive digital abstinence. Rather than being internalized, these components of well-being remain largely external and require ongoing maintenance through frequent contact. Participants reported the sense of all three being lost or at risk. However, the loss of belongingness was generally remarked on in more than one of the hourly entries made by hundreds of participants. Participants used five coping strategies to deal with the threat to their feelings of well-being.

Coping through Minimal Non-Compliance

This research demonstrates that digital abstinence directly affects participants' sense of belonging on multiple levels and, thereby, threatens their overall feelings of well-being.

The most popular coping mechanisms for protecting themselves from the subsequently negative state-of-being involved three forms of non-compliance. A small number chose to nap despite this being in violation of the twelve waking hours rule (7%). Naps often occurred before the sixth hour. This form of passive non-compliance enabled participants to take a psychological "time out" from their escalating displacement, anxiety and depression. In a sense they "reset" their psychological state. Consequently, they did not give up; after napping, participants resumed digital abstinence.

Far more participants used active non-compliance coping strategies. Twice as many participants (14%) tried to circumvent the point of being offline by getting a friend to check messages or social media sites and report back. This was typically a one-time behavior, after which abstinence was resumed. Fully one-quarter (25%) of participants were unable to resist and used their phones to check messaging and social media sites one to three times during the twelve hours.

Readers may be tempted to dismiss the experience of abstinence for these participants in view of their non-compliance, especially in the two active non-compliance examples. Consider, however, three-quarters of Millennials generally check their phones 50 to 100 times per day (Globe Newswire 2017). Checking three times in a twelve-hour period is still significantly abstinent. This explains the otherwise puzzling and seemingly self-contradicting stance of those who reported non-compliance and yet congratulated themselves on their fortitude during the twelve-hour experience. It is similar to someone taking pride in a week-long fast despite eating that one serving of avocado toast.

Coping by Changing the Behavior of Others

More than one-third of the participants tried to lessen their marginalized status by successfully changing the digital behavior of companions (family, friends, and intimates). Having companions share their physical-only presence provides a means to lessen the psychological burden of digital abstinence. Participants sometimes offered the rationale of ensuring their own success in completing the abstinence period by removing the constant temptation of seeing others digitally engaged. Abstinent companions also had the effect of minimizing the visibility of participant non-belonging because others in immediate proximity were also digitally disengaged. Further, it diminished the loss of centrality experienced when co-present companions remain digitally occupied while attending comparatively less to the physically present participant.

I can't do it so let's all not do it—getting others to "join in"

A number of participants report attempting to get companions to "join in" the digital abstinence with them but only a few succeeded (11%). Even so, most of this small group shared company for a relatively short period of time and got mutual digital abstinence for only part of that time.

> To help me with this assignment. I integrated my whole family in it with me [They only joined in for a few hours].
>
> While at the mall, I asked my girlfriend to participate in this exercise with me, which included mainly refraining from using her mobile phone. Being at the mall, I can tell she was getting frustrated not connecting with her friends.

If I can't do it, you can't do it—"imposing" digital abstinence

One-quarter (25%) approached companions more aggressively by attempting to impose restrictions on others:

> I didn't want to feel tempted to use my computer, so I hid the computer from my little brother.
>
> He wanted to use his iPhone and I told him he couldn't because I couldn't use the Internet so he had to sacrifice his iPhone also.
>
> I bugged my roommate while she tried to catch up on her bible readings on her iPhone. Since I was talking to her, being silly by making faces, and making up random songs, it took her twice as long to finish her readings.

Social Integration and Collective Consciousness

This research shows that digital abstinence diminishes a sense of well-being for Millennials by restricting affiliation, limiting expressions of emotional intimacy, and undermining their sense of belonging. Belongingness is a deep-seated need that can act as a strong motivational force. This could, in itself, explain participant cravings for connectivity; however, journal entries and participant reflections tell a more complex story.

Millennials and Collective Consciousness

It is clear that abstinence participants felt significantly distanced from their social groups. Further, they expressed feeling unattached and separated from their cohort as a whole.

> When I could finally be reconnected to the system or network that our generation considers a necessity I knew I would feel like me again.
>
> Most of the people my age have never lived without the Internet … It's no wonder why I felt so isolated, for that short time span.
>
> Due to the fact that a majority of people in my culture are online, or connected to digital media, it's no wonder I was taken by the overwhelming sense of being left out.

Millennals are well aware of themselves as a cohort. American culture has always categorized, labeled, and attributed particular characteristics to age-graded cohorts. This is especially strong in the case of Millennials who grew up alongside, and became strongly bound to and identified with, definitive technological change. The participants sometimes acknowledge this by referring to themselves as "digital natives." This cohort is largely responsible for establishing the social expectations that apply to digital behavior. Values, beliefs, and norms manifested in digital space have been, to an unprecedented degree, shaped by emerging adult and younger adult Millennials who provided the general expectations that socialized older users. This reverse-age socialization is supported as long as cohort members remain disproportionately early adopters and intensive users. It also reinforces the *collective consciousness* among Millennials, at least in terms of digital life.

Collective consciousness is a shared way of seeing and behaving in the world. It provides the framework for *social integration*—that sense of connection to one another and to the social setting.

> Without my phone, computer, or anything else … I felt like I was lost and was going to go crazy … when it was taken away from me I felt like I was missing something. It was like I was not connected to something and not involved in what was going on around me.

Routines and Mediated Life as Meaningful Life

Social integration allows people to make sense of their environment, their place in that environment, and the choices they make. The sense of connection and social integration is maintained through everyday normative activities in the form of routines, and in observing personal rituals as well as broader social rituals. The importance of routines was evident in numerous journal entries.

Scholars in philosophy and the social and behavioral sciences have long struggled over concepts about what makes life "meaningful." The focus in this chapter is different, and is based more on how individuals typically find meaning in their day-to-day existence. As such, the author has drawn on Heintzelman and King whose research demonstrates that people strongly associate simple concrete experiences with a meaningful life. Such experiences include "social relationships, the experience of pleasure, and the detection of reliable pattern or coherence in the environment" (2014:571). Routines enable a feeling of groundedness, coherence, and reliability. This includes quite basic activities. According to research, "regular appointments, morning coffee, and our typical commute … have been shown to affect, enhance, and support meaning in life" (Heintzelman and King 2014:564).

From this perspective, many find that their routines provide the foundation for feeling that they live a meaningful life. Participant entries provide support for this approach.

> It felt kind of weird not being able to share on social media that I was playing Monopoly with my daughter. I guess the norm of posting everything on social media is meaningful in my everyday life.

Participants in the current study expressed distress at forgoing the routines that intertwined their digital and physical presence, and provided a feeling of normalcy.

> During this hour I got ready to head for the gym and had breakfast. Because I couldn't do my regular routine, I felt as if I was punished.
>
> I am really irritated trying to figure out something I can do. It really irks me that my whole regular routine has to be thrown off; I feel unbalanced ....
>
> This a routine for me and I wasn't able to follow my routine, which bothered me ... I was just in a grumpy mood all day long.
>
> I did not know exactly what to do except that I had to relearn how to live, even when it came to regular daily routines.

Considering that social media postings are a regular part of participants' routines, it is not surprising that digital abstinence evoked feelings such as being irked, unbalanced, thrown off, bothered, and grumpy. Giving up routine social media and messaging applications constitutes digital social exclusion. Earlier studies indicate that "social exclusion reliably leads to lower ratings on meaningful existence ... even when this exclusion is momentary and impersonal" (Zadro, Williams, and Richardson 2004: 1). On the flip side, participants depicted their routine social media postings as extremely meaningful:

> The bright side is the fact that we all leave behind a digital footprint; [technology] allows us to connect in ways like never before ... Leaving a digital footprint behind literally allows us to create history and forever be remembered.

Marginalized within the Collective Consciousness

The power of collective conscious to promote feelings of social integration can be judged, in part, by the reactions of those forced to violate shared norms and behavioral expectations. Millennials who gave up their digital presence for half-a-day did not do so because they rejected the norms, behaviors, and understandings of their cohort. The fact that they remain part of the collective consciousness, while being unable to undertake the routine activities that reinforce and express it, resulted in three very clear patterns. First, they responded to this with feelings of shame and embarrassment. Second, they frequently indicated that their behaviors were non-normative and assumed others saw them as deviant. Third, they reported feeling alienated from their normal activities when experiencing them without a digital component.

The Shame of Maintaining Only a Physical Presence

Society is best able to maintain equilibrium when its members share and conform to a set of social expectations. Individuals who *internalize*—deeply accept, the importance of the values and norms associated with these expectations—will feel shame and embarrassment if their actions violate them. Journal entries frequently demonstrated these emotions. This was true even if the particular experience was positive:

> I also gave up an invitation to dinner because I knew we would be taking pictures and I did not feel comfortable going without my phone.
>
> Although I enjoyed my class so much more without these technologies, I could not help but feel ashamed.
>
> I wanted his phone number. If it was as usual, I would have taken out my phone and let him type his numbers in my phone. I felt bad, sorry for him, and little ashamed. I had to make him write it down.

Conforming is also a means of avoiding negative social judgment or being labeled a deviant.

Digital Abstinence as Millennial Deviance

The strength of collective consciousness can be seen when shared norms, are violated. Millennials forgoing digital connectivity saw themselves as looking deviant in the eyes of others. They described feelings of being judged.

> I felt kind of embarrassed using the payphone. I thought people were giving me weird looks while I was standing there talking to my classmate. I'm sure I was just paranoid but still, it made me feel uncomfortable.

> Inside [Starbucks] I got the sudden urge to use my phone … I began to feel a sudden anxiety to use my phone or at least act as if I was using it.
>
> Walking around campus just observing everyone and how technology is important to all. I saw everyone on their phones and laptops with their earphones on. I couldn't do that but I felt like I needed to in order to not look awkward around other people.
>
> I felt like a loser telling her I didn't have a cell phone [with me] … It was a very weird feeling because she looked at me like there must have been something really wrong with me ….

If digital life was comprised of a set of activities, without deeper social meaning, avoiding digital presence would not be experienced as shameful, deviant, or potentially worthy of negative social judgment.

Incompleteness and Dissatisfaction

Social integration depends not only on the shared norms and values of collective consciousness, but also on value-based behavior. For participants, the loss of digital connectedness undermined feelings of social integration when digital components were omitted from their regular activities. Consequently, such activities seemed less meaningful, and participants became alienated from their own actions. Journal entries described significant emotional or sensory disconnects from day-to-day experiences, even meals and showers.

> I hate this feeling of restraint … not being able to multi-task. I feel disconnected from the outside world, not being able to interact and socialize through the net. Instead I began eating while the food became tasteless.
>
> My breakfast was delicious, but it didn't taste the same because … I can't watch any YouTube videos while eating.
>
> I was not happy with the outcome because I take morning showers to get energized. If this affect is absent [without Pandora], then taking a morning shower is worthless.
>
> I felt that the time to take shower was really boring, empty and long.
>
> I started to feel depressed because I haven't seen my house this quiet before. Without technology, my house was boring and it felt dead. I had to get out of there.

Feeling that commonplace activity is diminished without some form of digital presence is not related to solitude. Participants' reactions remained consistent, whether or not they were accompanied by family, close friends, or intimate partners. This consistency indicates that co-presence—that is being simultaneously in physical and in digital space—is a deeply-internalized norm.

> When I'm at a restaurant waiting to be seated, waiting to order, or waiting for my food, I am usually on my phone. This time, I was not able to do that, so at times my mood would be anxious, bored or irritated about giving it up. Instead, I focused on my boyfriend and our conversation.
>
> Even as we sat and waited for our food, there would be times where we had a moment of silence. As if we had nothing to talk about and our lives seemed bleak at that moment.

Dissatisfaction with only the physical presence aspect of day-to-day activities was expressed by more than a third of the participants (38%).

In addition to the sense that specific tasks were less satisfying, participants frequently expressed distress about being unable to control their immediate context. Their use of language signaled heightened awareness of physical-only presence as a constraint. Participants spoke of: being "forced" to experience silence; having to "actually listen;" being "unable" to tune out environmental stimuli; "having" to mono-task; being "required" to pay more attention to class lectures, driving, navigation, and strangers; and "having" to experience something directly, while being "forbidden" to digitally validate it. For example:

> Instead of using my cell phone, I had to … **actually listen** to the guest speakers.
>
> While at the mall, with my girlfriend, I was getting frustrated not connecting with my friends … we are **confined to each other's company**, without any email or text message interruptions.
>
> I wanted to take pictures of our contest but couldn't which sucked! I felt sad … **it forced me to enjoy the moment** without the option of sharing it.

Overall these expressions of dissatisfaction indicate another disconnect from feelings of social integration. Being restricted to only a physical presence deeply challenges Millennial perceptions of their environment and how they fit into that environment.

Alienation as Inertia and Moodiness

Participants felt alienated, not only from routines, common tasks, and their social settings, but also from themselves. Two out of every eleven participants described themselves as not feeling whole. More than one-third of the participants (36%) lost all motivation and were unable to think of anything to do for significant periods of time.

> I felt sluggish with no one to interact with [online]. I didn't feel like my complete self. I was unable to talk to customers in a friendly manner. Many customers didn't feel my energy and sincerity and left the store empty-handed, causing my team to lose sales.
>
> If we are prohibited from using some type of technology we feel anxious, impatient, incompetent, almost handicapped.
>
> I felt gloomy [without YouTube]. It just didn't feel right. I wasn't complete … I was sitting on my bed, not wanting to get up, time was passing by slowly and I didn't want to go to work.
>
> I lay on my bed for what was like forever. I was just lying there lifeless … I couldn't wait [to be online] and I couldn't take it any longer. So I lay there, just lay there. I didn't go to sleep but just lay there.

Others exhibited strong emotions or reported mood swings. In a single hour, participants might report surprise or happiness in some formerly unnoticed or undervalued quality in the directly-inhabited environment (27%) which dissolved into anger and frustration (21%) or sadness and depression (17%) in having to sacrifice control over, or delay mediating, some aspect of that same environment. Similarly, many reported appreciating an in-person interaction with family (27%), friends and intimates (15%), and even strangers (10%) that they may not have experienced otherwise, then moved on to express an intense craving for digitally-mediated interaction, messaging, and social media. One participant expresses it well:

> My mood is a roller coaster. I find something to do and my mood is happy and content, but as soon as I complete my task, I feel irritable. What's wrong with me?

Many participants ended their journals with their sentiments on being able to resume co-presence. It is telling that they focus on redressing their overall deprivation rather than any specific activity.

> I feel great. In just a few minutes I will be connected to the social media and I will feel fully alive.
>
> I am anxiously awaiting [the 12 hours to end] so I can get back to the "real world" … and use current technology.

Conclusion

Digital life is not as simple as it is often conceived. It is not just a set of analogs to physical experiences. It is not simply a set of online behavioral activities that can be juxtaposed to offline activities. It has become part-and-parcel of who we are in our daily lives and increasingly interpenetrates our directly inhabited, physical world. It is meaningless to continue to think in binary terms about offline (physical presence) and online (digital presence), because both interweave to become the co-presence of life as it is lived. This is why reactions to deprivation were so extreme. The importance is not in just the behaviors that arose, but in how co-presence addresses basic needs for a sense of well-being, indicates social integration, manifests collective consciousness, and contributes to the feeling of living a meaningful life.

References

Agger, Ben. 2012. *Oversharing Presentations of Self in the Internet Age*. New York: Routledge.

eMarketer. 2017. "Understanding Teens and their Smartphone Habits." Mobile Demographics, September 5. Retrieved July 13, 2018 (https://www.emarketer.com/Article/Understanding-Teens-Their-Smartphone-Habits/1016423).

Erber, Ralph and Maureen Wang Erber. 2001. *Intimate Relationships: Issues, Theories and Research*. Boston: Allyn and Bacon.

Gannon, Megan. 2013. "Why Some Facebook Users Constantly Update Status." Live Science, Jan 3. Retrieved Jan 1, 2014 (https://www.livescience.com/25972-facebook-status-updates-loneliness.html).

Globe Newswire. 2017. "Smartphone Obsession Grows with 25% of Millennials Spending More than Five Hours Per Day on the Phone." May 18. Retrieved June 26, 2018 (https://globenewswire.com/news-release/2017/05/18/987607/0/en/Smartphone-Obsession-Grows-with-25-of-Millennials-Spending-More-Than-5-Hours-Per-Day-on-the-Phone.html).

Heintzelman, Samantha and Laura A. King. 2014. "Life is Pretty Meaningful." *American Psychologist* 69(6):561–574

Jiang, JingJing. 2018. "Millennials Stand Out for Their Technology Use, But Older Generations Also Embrace Digital Life." Pew Research Center, May 2. Retrieved June 22, 2018 (http://www.pewresearch.org/fact-tank/2018/05/02/millennials-stand-out-for-their-technology-use-but-older-generations-also-embrace-digital-life/).

Miller, Vincent. 2011. *Understanding Digital Culture*. Thousand Oaks, CA: Sage.

Perrin, Andrew and JingJing Jiang. 2018. "About a Quarter of US Adults Say They Are 'Almost Constantly' Online." Pew Research Center, March 14. Retrieved June 22, 2018 (http://www.pewresearch.org/fact-tank/2018/03/14/about-a-quarter-of-americans-report-going-online-almost-constantly/).

Richter, Felix. 2018. "Digital Detox: America's Smartphone Addiction." Statista, June 20. Retrieved July 13, 2018 (https://www.statista.com/chart/12403/smartphone-addiction/).

Smith, Aaron and Monica Anderson. 2018. "Social Media Use in 2018." Pew Research Center, March 1. Retrieved June 22, 2018 (http://www.pewinternet.org/2018/03/01/social-media-use-in-2018/).

Statista. 2018. "Reach of Selected Social Networks in the United States as of February 2017, by Age Group." Retrieved May 10, 2018 (https://www.statista.com/statistics/305245/us-social-network-penetration-age-group/).

Turkle, Sherry. 2011. *Alone Together: Why We Expect More from Technology and Less from Each Other*. New York: Basic Books.

Zadro, Lisa, Kipling D. Williams and Rick Richardson. 2004. "How Low Can You Go? Ostracism by a Computer Is Sufficient to Lower Self-Reported Levels of Belonging, Control, Self-Esteem, and Meaningful Existence." *Journal of Experimental Social Psychology* 40(4):560–567.

2

DIGITAL INK

Social Media and Tattoo Culture in Consideration of Gender

Deborah Louise Burns

Tattoo Popularity

There is no doubt that tattoos have become popular among Millennials around the world, especially in the United States. According to a Pew research study, nearly 38 percent of Millennials have at least one tattoo, compared to 32 percent of Gen Xers (Pew Research Center 2010). Even though tattoos and tattoo culture are vastly popular in this millennial era, historically this has not always been the case. Since its reintroduction to the upper classes in Europe and the US working class in the eighteenth century, modern tattoos and tattoo culture have had a remarkable journey. In just over a century, societal beliefs shifted from deviance to acceptance, and viewpoints continue to change. Once considered the mark of criminals, deviants, and other marginalized groups, perceptions have changed during the last 40 years in the US, and on a global level. This is due, in part, to increased exposure to tattoos through mass media in mainstream culture. In the current social climate of fashion trends and self-expression, the once unacceptable tattoos are now everywhere. In the millennial age audiences now applaud "actors, models, musicians, and professional athletes as they proudly herald the mainstreaming of a previously marginalized and historically underground practice" (Kosut 2006a:1035).

From newspaper and magazine articles, to television shows, tattoos are everywhere. The prevalence of tattoos on social media and social networking sites (SNS) including Twitter, Instagram, Pinterest, and Facebook, along with tattoo-related stories on various mainstream Internet news and entertainment outlets like Buzzfeed, constitute undeniable evidence that tattoos have become a fixture in American mainstream culture. This is thanks to a strong digital presence in social media that connects to a vast audience.

This chapter explores the role that social media has had on US tattoo culture. Considerations of images from print magazines to digital sites indicate that cooptation by mainstream media has contributed to gender and beauty standards presented in the virtual world. It has also affected disclosure among the members of tattoo culture. In order to understand the changes in tattoo culture, and how it is being affected by online representation and interaction, a brief historical overview is necessary. Consequently, the following discusses the rise of tattoo culture in mainstream media, the prevalence of tattoo culture online, and how its social media site representation has impacted online social interactions. A content analysis (see Chapter 15) was conducted on the images and videos posted on the *Inked* Facebook web page in order to examine possible patterns in the way gender is presented online, and to see if these are the same gendered patterns that have been featured in tattoo magazines—an important media source that has played a significant role in the mainstreaming of tattoo culture.

Tattoos in Changing Historical Context

Contemporary tattoo culture and tattoo images developed and evolved in a social context. Unfolding this context contributes to an understanding of the way tattoo culture became a fixture in the digital world. Tattoos have been a part of human society for centuries. Initially, various groups used them to symbolize membership and rites of passage (Barron 2013; Sanders 2008). However, since tattoos were reintroduced to the Western world in the late eighteenth century, their meaning and purpose transformed. This is especially evident in countries like the United States, where views on tattooing have fluctuated among perceived deviance, to accepted self-expression, pride, and artistry throughout various periods of time. As Professor Mary Kosut noted: "Tattooing has undergone many voyages and has crossed continents, cut across class and gender lines, and drifted between high and low cultural backgrounds" (2006b:74).

Although the *practice* of tattooing has been around for centuries, the associations people held about those who *have tattoos* shifted, in relatively-modern history, among exotic commodity, membership symbol, fashion fad, and deviant practice. During their introduction, tattoos gained favor and achieved a degree of mainstream popularity among the upper class in Europe, and the working class in the United States—especially among members of various branches of the US military (Adams 2009; Kosut 2006b; Fisher 2002; DeMello 2000). As the popularity of tattoos grew, the increasing demand for tattoos led to the advent of the tattoo machine by Samuel O'Reilly in 1891, which gave rise to the foundation of the modern tattoo industry (Barron 2017; DeMello 2000; Mifflin 2013; Sanders 2008). The tattoo machine paved the way for professional tattoo artists and allowed the new-found industry and culture to flourish as various mainstream groups—such as the working class and members of the military—and marginalized groups—like carnies and gang members, marked their bodies to express their sense of adventure, group membership, pride, and patriotism (Kosut 2006b).

Losing Allure and Gaining Stigma

Although the advent of the tattoo machine and creation of the tattoo industry had some positive impacts on tattoo culture, unforeseen consequences that occurred with these events had less favorable impacts on the initial popularity of tattoos. The tattoo machine not only made the application of tattoos faster and less painful compared to the traditional hand-tapping method, but it made tattoos cheaper, which made it a more accessible art form for groups that were considered less desirable or marginalized (DeMello 2000). As tattoos became more widespread and affordable, marginalized groups adopted them as symbols of group membership and to express rebellion against mainstream society. This increased the popularity of tattoos among some populations while reducing it in other populations. They lost their adventurous and exotic appeal to the upper classes (Fisher 2002; DeMello 2000). As even more members from marginalized groups received tattoos, they became more strongly associated with deviance and served to label, and further marginalize, those members who were tattooed. This was true in both Europe and the United States. This widespread practice of tattooing among the lower class and social deviants led to the abandonment of tattooing by the European elite by the beginning of World War I, although it remained popular among those in the American working class until the end of World War II (Fisher 2002; DeMello 2000).

By the middle of twentieth century, the attitudes towards tattoos and its culture in the US began to change. After World War II, the popularity of tattooing began to fade, and the rise of the new middle class saw a marked shift on views about tattoos. The general outlook began to move from tattoos as acceptable to deviant. As subcultural groups— such as bikers, convicts, and Chicano gang members—used tattoos to signify membership, they contributed to the stigmatization of tattoo culture (DeMello, 2000). During this time, tattoos became a "form of deviance and a challenge to both emerging mainstream middle-class values as well to traditional forms of patriotic and love-inspired working-class tattoos" (DeMello 2000:67). This pattern was repeated as those who were considered marginal and deviants continued to adopt tattoos for various reasons. The practice, as a whole, came to be frowned upon by the wider society, leading to the belief that tattoos were the marks of criminals and outcasts; they were no longer viewed as normative within mainstream culture.

Tattoo Stigma and Psycho-Biology

The connection between tattooing and deviance was significantly strengthened in the United States after theories developed by European experts in criminology and the medical field conducted research on deviance, where they reported a connection between deviance and tattoos as resulting from psychobiological factors. In other words, the motivation for getting tattoos was seen as part of the hereditary personality traits criminologists of that era believed led to criminality (Barron 2017; Barron 2013; Adams 2009; Kosut 2006b). Tattoos were the focus of much research during the twentieth century; however, most of these primarily focused on those who were incarcerated or in mental institutions. This time period pre-dates mass incarceration, so this was only a small portion of the US population (Adams 2009; DeMello 2000). Further, the nature of the sample was biased at the outset by focusing on inmates. Nevertheless, those studies were generalized to the US population, giving those who were tattooed—regardless of their background or group affiliations—the label of "deviant," and reinforced the presumed correlation between deviance and tattoos (Adams 2009). Unfortunately, these theories became widely popular and accepted by many, influencing and justifying their negative attitudes towards tattooed individuals. These attitudes created a social stigma—that is, a generalized strong disapproval—that attached to all those who were tattooed. Even when theories on the biological connection between tattoos and deviance were later "debunked, their influence had clearly resonated in popular and scientific conceptions of tattoos," in part because it coincided with changes in which social groups were getting tattoos, and in what motivated them (Kosut 2006b:80).

Along with the socio-biological and medical perspectives on tattooing, the increasing association of tattoos with marginal social groups cemented their status as indicators of deviance. By the mid-twentieth century, tattooing was widely believed to be a deviant practice by mainstream society, stigmatizing for many decades those who had already gotten tattoos (Adams 2009; DeMello 2000; Sanders 2008). This became problematic for tattooed individuals, whether or not they were a part of a deviant group. This stigma was not because their behavior violated social norms, but because their tattoos made others *assume* they had negative personal or social characteristics and should, therefore, be excluded (Goffman 1963). Being labeled as deviant not only limited their social networks, it also made them relatively difficult to employ.

From Deviant Ink to Art

The 1960s ushered in an emphasis on the arts. Attitudes towards tattoos and tattoo culture shifted with the emergence of the *Tattoo Renaissance*—a period marked by technological, artistic and social changes, where tattoo artists began to change "as young men and women with fine art degrees began to enter the profession alongside older, traditional trained tattooists" (DeMello 2000:72). The new generation of tattoo artists during this period introduced a new style of tattoos that differed from the style associated with deviance. This played a role in changing the attitudes about tattoos in US mainstream culture, especially among members of the middle class. "These newer tattooists took their artistic inspiration from other cultures such as Japan, Borneo, Samoa, and Native America instead of traditional North American designs and in this process began to appeal to a middle-class audience" (DeMello 2000:72). Simultaneously, mainstream media played a crucial role in changing attitudes towards tattoos. As the media began to focus on tattoo culture, some tattoo artists took advantage of this newfound media interest. Tattoo artists in America began using reputable mainstream media to showcase tattooing as art. For example, "tattoo artist Lyle Tuttle and his shop were featured in popular magazines, local newspapers, and on television shows throughout the seventies, which stated tattooing's emerging popularity in mainstream culture" (DeMello 2000:76).

Tattoo Culture and the Media

The popularity of tattoos among the US middle-class continued into the eighties and the nineties as the exposure of tattoos in the mass media increased. It was during the eighties and the nineties that tattoos as both an art form and a social practice were transformed through media exposure. This had a profound influence on the tattoo culture and on those who did not have tattoos. Americans were exposed to images of tattoos as they read popular magazines and local newspapers. This "distinguished a new tattooing community from its origins, through highly selective and standardized presentations of how tattooed people were like" (DeMello 2000:98). US mass media began to portray tattoo artists and their customers as highly educated, which connected the middle and upper classes with tattoos. The media began to portray tattoos to fit middle-class taste by placing an emphasis on white-collar professionals who were getting tattooed by "prestigious tattoo artists who held a Master of Fine Arts degree" (DeMello 2000:100). This type of portrayal of tattoos in the media allowed for the mainstream culture to "take-over" aspects of tattoo culture—coopting it—while redefining it as a product to capitalize upon—resulting in commodification (Kosut 2006a).

As tattooing received more positive US mainstream exposure through mass media outlets, it broadened audience interest in the topic. This led to the creation of tattoo magazines, which also contributed to the shift of tattooing perceptions among the mainstream from deviance to acceptance. Throughout the eighties and nineties, a number of tattoo magazines were introduced to the public. Some tattoo magazines, like *TattooTime*, had a very short life span, while many others like *Tattoo* survived and are still popular among tattoo professionals and tattoo wearers, so much so that they are now readily available at newsstands, bookstores, and local grocery stores. With the advent of the digital age, some even interfaced themselves through social media platforms such as the pages of Facebook. Whether these tattoo magazines had a short life span or stood the test of time—all of them had a profound impact on tattoo culture and how it came to be perceived by mainstream culture. "Tattoo magazines are influential because many are produced by fellow tattoo artists and tattoo wearers, which carries a much greater authority than publications produced by people who are not members of the tattoo community" (DeMello 2000:101).

When tattoo magazines were first introduced to the public, they were not all viewed in the same way, and the presentation of tattoos and the culture associated with them affected the way they were viewed. As the publication of tattoo magazines became more prevalent during the eighties and the nineties, two different categories of tattoo magazines, lowbrow and highbrow, were being disseminated. These designations into lowbrow and highbrow were based on the editorial content of the tattoo magazines, which was determined by the publisher to cater to the target audience's taste (Thompson 2015; Mifflin 2013; DeMello 2000).

Highbrow and Lowbrow

Highbrow tattoo magazines were published in the eighties and the nineties in order to introduce tattoos to a broader audience. Highbrow tattoo magazines focused on "legitimatizing tattoos and its culture by investing it with middle class cultural capital by having a heavy reliance on references to fine arts, academic disciplines, and to mainstream journalism" (DeMello 2000:101). Highbrow tattoo magazines presented articles that focused on high-profile tattoo artists and their middle-class clientele, while excluding any images of non-middle-class tattoo wearers. They shaped their presentations in order to distance themselves from the lowbrow tattoo magazines that were also available to the public (DeMello 2000).

In contrast, lowbrow tattoo magazines tried to set themselves apart from their highbrow counterparts by taking control of the perception of tattoos in mainstream culture. Publishers tried to present their lowbrow tattoo magazines as providing:

> A more authentic representation of the tattoo culture as they attempted to insert their points of view on tattoos in mainstream culture; however, unlike highbrow tattoo magazines, lowbrow tattoo magazines utilized more sexually explicit photos, more working and lower-class images, have a devoted fan section, and

> showcased advertisements from tattoo suppliers and adult entertainment companies (DeMello 2000:113).

Lowbrow tattoo magazines focused more on the alternative world of tattooing that was inclusive of the working and lower classes as it showcased lifestyles that did not appeal to the middle-class.

The introduction of these magazines propelled tattoos into US mainstream culture and contributed to the shift in the view on tattoos, from a deviant to a more tolerable and acceptable view throughout the eighties and nineties. However, the two categories of tattoo magazines also showcased a class difference and stratification in the tattoo culture (DeMello 2000).

Broadcasting Ink

The dissemination of tattoos and depictions of the culture in mainstream media during the seventies, eighties, and nineties has allowed for a large audience to view tattooing in a light not seen as deviant. The exposure and integration of tattooing into mass media over those three decades, once more, made tattooing acceptable and popular in the US. By the beginning of the twenty-first century, American television and cable networks capitalized on this new-found popularity. Television and cable networks introduced several tattoo-focused shows, as tattooing continued to be a subject of fascination in US mainstream culture. Multiple television specials, documentaries, and reality television shows dedicated to tattooing aired on various cable networks. Those without access to cable television were able to watch these shows through video sharing websites and on-demand Internet streaming media services. This explosion and saturation of the tattoo culture on television has been one of the biggest influences in the shift in the US mainstream view on tattoos.

During this period, cable networks began airing documentaries and shows on tattooing that explained the process and the reasons behind getting tattooed, such as *Skin Deep* (2001) and *Tattoo: Under the Skin* (2005). These educational shows continued throughout the first part of the twenty-first century with popular specials such as *Ancient Ink* (2008) and *Tattoo Nation* (2013). Documentaries on tattoo culture disseminated the message to the public that tattoos were artistic, meaningful, sophisticated, and spiritual for people of all backgrounds, especially for those from the middle and upper classes.

Along with these specials and documentaries, the twenty-first century also saw the birth and growth of reality television shows based on tattoo culture. For example, in 2005, America was introduced to the reality television show *Miami Ink*, which led to several spin offs including *LA Ink* (2007) and *NY Ink* (2011). Along with *Miami Ink*, other networks developed their own tattoo-based reality television shows such as *Inked* (2005) and *Tattoo Highway* (2009). There was, however, a growing concern about how authentic these shows were when it came to representing the tattoo culture and the lives of tattoo artists and tattoo wearers, even as tattoo-based reality television shows kept growing in popularity. Even though the artificiality of these reality shows came into question, their sheer popularity maintained a demand for these shows as television and cable networks kept producing new tattoo reality television shows such as *Tattoo's After Dark* (2014), *Black Ink Crew* (2012) and *Black Ink Crew Chicago* (2015), *Tattoo Nightmares* (2012), *America's Worse Tattoos* (2012), and *Bad Ink* (2013). Even competition-based tattoo reality shows such as *Ink Master* (2012) and *Best Ink* (2012) were created. Both tattoo competition shows are affiliated with current popular US tattoo magazines, *Inked* and *Tattoo* respectively.

The continued use of the media, including popular tattoo magazines and television shows, helped change the perception of tattoos by highlighting its culture and bringing the practice out from the underground scene and into the mainstream (Kosut 2006a; Kosut 2006b). By the beginning of the millennial era, tattoos had become a visible part of the public sphere as tattoos and tattoo culture are continuously featured in the mass media; however, the introduction and use of SNS changed the way that people learn about and participate in tattoo culture.

Social Media and Tattoo Culture

Tattoos are more widespread than ever thanks to the Internet and SNS, such as Facebook, Instagram, and Pinterest. SNS have contributed to the popularity of tattoos and its culture in a unique way. Unlike television shows and other media formats, SNS has allowed a way for tattoo aficionados and novices to view and interact with other members of the tattoo community. In other words, the exposure of tattoos in SNS has influenced the socialization of the viewers, increased commodification, and accelerated cooptation of the tattoo by mainstream culture.

In general, SNS has changed the way people interact with others by changing the way people and groups connect, and how they present themselves in digital space to interact and communicate. SNS has allowed people to communicate with a vast audience who have a shared interest in a digital environment. This has allowed people to reach out and meet new individuals, become a part of digital communities, promote their interest, and present themselves or topics through digital platforms that would have been difficult or impossible in a face-to-face setting. Because of this, SNS has played an important role in the spread of tattoos and its culture as it has created a unique way for people to interact and socialize with one another. Social media platforms such as Facebook have given tattoo artists, and entire shops, the ability to promote their work and reputation to an immense audience. SNS has also allowed members of the tattoo culture to form digital communities through social media platforms. Some tattoo artists have created online communities like Tattoodo, and claim that they are the world's largest tattoo community as it is a one-stop hub for tattoo culture. Lady Tattooers is another such community, dedicated to showcasing the work of female tattoo artists around the world. Facebook sites like these promote community and culture by sharing tattoo photos and stories and asking for audience participation in the form of comments or responses of "like," "love," "haha," "wow," "sad," or "angry." This allows the audience to interact with others on these pages and builds a sense of community for fans and newcomers.

Along those lines, many popular tattoo magazines also began using social media platforms such as Facebook to form a digital community with the idea of promoting tattoo culture and continuing the growing popularity and attitude change towards tattoos. Similar to other online communities, these digital tattoo magazines have a vast following in digital space, which is something that could not be achieved in the printed format of physical space. For example, according to their Facebook page, *Inked* Magazine has over 11 million followers, *Tattoo* Magazine has over 2 million followers, and *Skin and Ink* Magazine has a little over 50 thousand followers. Tattoo magazines like *Inked* and *Tattoo* have also used Facebook pages, not only to present images and content from their magazine, but to also encourage their audience to like their page, in order to view all of their posts in their Facebook feed.

These tattoo magazines with Facebook pages are interactive with their audience and encourage audience participation and interaction through their posts and live streaming video sessions that involve real time question and answer sessions.

Online Patterns and Gendered Images

Although interaction does create community in a digital environment, there are several issues with the promotion and expectation of audience participation in communities created by these magazines. One issue is how the mainstreaming or cooptation of the tattoo and its culture by the media has contributed to specific gender and beauty standards online, and how that has affected disclosure among its members. In order to explore this, images in both posts and videos from the *Inked* magazine Facebook page were analyzed to see if gendered images were being presented to the audience and, if so, whether audience participation seemed to be influenced by these images.

The Facebook page of *Inked* magazine was chosen because it is a popular highbrow tattoo magazine, with one of the largest followings compared to other tattoo magazines on Facebook. This analysis on gendered patterns in online representation and self-presentation shows similar patterns to those found in a previous research study. The findings from a content analysis on tattoo magazines showed support for the hypothesis that popular tattoo magazines presented gendered images, where women were "presented with an emphasis on physical appearance and titillation or hypersexualization and that men were presented with a more neutral tattoo-as-art emphasis" (Burns 2014).

After analyzing *Inked*'s Facebook page, the patterns captured in the content analysis showed that the page was in line with the findings from other studies that looked at presentation of gender. When analyzing the posts that only contained photos on *Inked*'s Facebook pages, three-quarters (75%) of these posts were of images of women, while only one-quarter (25%) were of men. Less than half (42%) of these posts focused on art, while over half (58%) focused on titillation. Most (87%) of the images of men were presented in a more neutral tattoo-as-art emphasis, while the majority (73%) of the images of women were presented with an emphasis on physical appearance and titillation. In the case of the images of men, the focus was on the men and their tattoos, without the need to present their entire body. Over a quarter (27%) dressed in modest clothing and another third (33%) dressed in revealing clothing. On the other hand, in the majority (76%) of the images of women in these posts, were dressed in either revealing clothing (53%) or in lingerie (23%), and the majority of these images had a mainstream body type (86%). Several of these posts that presented women in an objectified manner and used suggestive wording with the image such as "No excuses" or "tats and tatas!" Similar to the findings from the previous tattoo magazine studies, as tattoos became more prevalent in various mainstream media outlets, those in control of these SNSs have chosen to present gender in the same way that mainstream magazines or television shows would "in order to capitalize on profits by objectifying women's physical appearance in order to attract viewers instead of focusing on the artwork of an individual's tattoo" (Burns 2014:48).

Although the mainstream acceptance and cooptation of tattoos and tattoo culture may have some positive affects for those who are members, one of the negative effects is how gender is presented and perceived. The tattoo and tattoo culture have been coopted or "repackaged" to make them less deviant and threatening for middle-class taste; at the same time cooptation has created other social problems such as the false need of the "classed body," and the watered-down meaning of tattoo culture and its practices (Burns 2014; Adams 2009; DeMello 2000; Bourdieu 1984; Hebdige 1979).

Some of the posts that are featured on the *Inked* Facebook page also involved and encouraged audience participation. For example, in 2014, *Inked* asked their fans: "Want to show off your ink here?" and encouraged the audience to send and share their selfies. This resulted in a 15-part series called "Facebook Fans of *Inked* Selfies" that was posted on both *Inked*'s website and their Facebook page in 2014 and a few other times, as recently as 2017; with thousands of reactions, comments, and shares. A separate content analysis on just the 2014 "Facebook Fans of *Inked* Selfies" showed almost all (98%) of the participants who had their submitted selfies selected were women; further, 83 percent of them presented themselves with an emphasis on physical appearance and titillation. This is problematic since these participants were presenting more than their tattoos in their selfies, and as participation like this brings up issues of disclosure and oversharing through their self-presentation. As Baym stated, "when we are disclosing to a group on SNS or to a community, our behavior is also visible to many other people" (2010:147). This can be perceived as inappropriate, and can have a negative impact on these participants that may lead to embarrassment, tarnished reputations, and the loss of employment and future opportunities (Trottier 2014; Agger 2012). This behavior may be judged by some as falling under the idea that "people should know better" or have "common sense;" however, this oversimplifies the dynamics behind such sharing. Since the act of sharing or submitting somewhat personalized posts has become a normalized or routine type of activity on SNS, many users forget that the act of sharing or submitting personal details, such as images of themselves, can lead to consequences that they do not initially expect or consider, such as the type of social stigma that can discredit and tarnish a person's reputation. This behavior can impact their social lives both online and face-to-face (Trottier 2014; Goffman 1963).

Along with images, video posts—including live stream—have recently become more common on the *Inked* Facebook page. Like image or photo-based posts, live stream videos involve audience participation and interaction; however, the interaction is done in real time. Many live stream videos involve real time activities where tattoo aficionados and novices can interact with those hosting the videos by typing back what they want to say in the comments section. Those who are featured in the videos, or the host, will ask the audience for their opinions and ask questions that can cause the audience to disclose information, such as where they are from—all this without the audience focusing on the fact that they are disclosing personal information.

When it comes to video posts versus other types of posts, one would expect to see a similar pattern of gendered presentation, where men are presented with a neutral tattoo-as-art emphasis, and women presented with an emphasis on physical appearance and titillation. However, unlike the still image posts, most of the videos focused on both men and women with a tattoo-as-art emphasis (72%). These fairly recent videos center around professional artists working on their clients, showing new tattoo work, activities at tattoo conventions, and what techniques the artists are using to apply the tattoos. The other video posts (28%), featured women in lingerie or revealing clothing and focused on titillation. Although this is a relatively more promising pattern, it is still disconcerting to consider that, over time, the videos on social media platforms may well change to align with patterns that have been seen in tattoo magazines and in photo posts on SNS platforms (Burns 2014).

Although many factors contribute to how the images presented on the *Inked*'s Facebook page are gendered, two main factors have major impact. These are: (1) the dominance of male-owned tattoo media; and (2) the cooptation of tattoos by mainstream culture. Both have played a crucial role in the objectification of women in tattoo culture. Magazine editors and publishers, television producers, and tattoo-oriented social media hosts are typically men. Tattooing as a profession is male-dominated. Most of the media revolving around the tattoo culture is, likewise, male dominated. It is not surprising that the media around tattoos and tattoo culture is male centric (Mifflin 2013). This can be seen as women are typically presented in tattoo-related media with an emphasis on their physical appearance, and in hypersexualized ways, as objects for men's pleasure—a common tactic in mainstream media (Hatton and Trautner 2011). Tattoo magazines are an example of this as these magazines present women "as scantily clad sex objects" with an emphasis on their bodies, while men are presented as professionals, supporting "traditional gender stereotypes" (Thompson 2015:201). There is some debate over the difference between highbrow and lowbrow tattoo magazines; however, early magazines, whether considered highbrow or lowbrow, had very little representation of women tattooists. This created an issue of invisibility and marginalization (Mifflin 2013). In the case of more current tattoo magazines, very little difference remains between what was considered to be highbrow and what was considered to be lowbrow (Burns 2014). This issue is also mirrored in the posts presented on the *Inked* Facebook page, as most of the women were presented as objects of pleasure, drawing audience focus away from women's tattoos and emphasizing the focus on their bodies.

Some may argue that the presentation of women emphasizing their bodies can be seen as an empowering and liberating way of reclaiming women's bodies; this can be a double-edged sword. Gendered stereotypes, objectification, hypersexualization, and internalized oppression can occur, resulting in negative and unhealthy consequences for those presented in media images, including those who are a part of the tattoo culture. Who has power and control over of which of these images are selected, what types of images are given more concentrated exposure, and the context in which they are presented in any media outlet, including SNSs, must also be considered. An image can be gendered and perpetuate traditional gender stereotypes, where it can objectify individuals and potentially stigmatize them. Saturation can perpetuate negative and sexist stereotypes for an entire community. For example, the emphasis on objectifying the bodies of tattooed women in the media, including SNSs, has been problematic in perpetuating the stereotype that labels tattooed women as sexually deviant or promiscuous (Santos 2009; Viren and Furnham 2007; Braunberger 2000).

Besides the factor of male-owned media, cooptation of tattoos by mainstream culture is another factor contributing to the issue of gendered images presented on *Inked*'s Facebook page. This can be seen by the way most of the women who were presented with an emphasis on titillation also had a body type that fit mainstream beauty and body ideals. Mainstream print and broadcast media is saturated with images that perpetuate very limited notions on "ideal" body types for women. This supports the negative effects of the cooptation of tattoos and its culture by mainstream culture. Many believed that tattoo culture was an alternative expression of beauty and allowed women to contest mainstream beauty standards, and much of this was supported by previous research (Atkinson 2002; Braunberger 2000). However, the cooptation of tattoo culture by mainstream culture has replaced the alternative beauty standards with what has long been considered acceptable or beautiful by mainstream standards.

The proliferation of such images that are unvarying and do not leave room for diverse body types can lead to issues of internalized oppression—where women come to believe, and act on, the perspective of female worth being linked to predominant, and very narrow, beauty ideals (Burns 2014). This was observed in most of the picture-focused posts, as they did not feature diverse bodies, but featured more mainstream body types. This was also observed with audience participation in the Facebook Fans of *Inked* Selfies series. Individuals perceived a need to present their overall physical appearance along with the tattoo in a selfie. This indicates the cooptation of tattoos from alternative to mainstream in the standards and expectations of beauty that now flood the tattoo subculture. This is problematic as "physical appearance has profound implications and can affect physical and emotional health of women as they internalize these images and create an unhealthy expectation and need to fit a beauty standard that is not obtainable for many people" (Zones 2005). The internalization of the false need for mainstream beauty and body type by tattooed individuals, now prevalent on social media, has created the rigid belief that tattooed women must fit that ideal to be considered worthy of sharing posts of their tattoos (Burns 2014; Bourdieu 1984).

Conclusion

Attitudes towards modern tattoos and tattoo culture have continuously fluctuated. Outlooks have swung between acceptable and deviant over most of tattoo history. At this point, however, the popularity of tattoos in mainstream culture is expected to keep growing along with its prevalence in mass media, especially SNSs. However, it is important to recognize the issues that come with mainstream cooptation and media representation, especially with representation of gender within SNSs. Studies like this one can bring awareness to these issues.

The findings from earlier research involving tattoo magazines, are replicated in the analysis of digital platform images of tattooed individuals on SNSs. Women were presented with emphasis on physical appearance within mainstream body-type ideals and titillation, whereas men were presented with more neutral tattoo-as-art emphases (Burns 2014). These issues need to be recognized, but they also need to be addressed. Studies like this one are important for highlighting the fact that gendered images in which women are objectified are problematic; they perpetuate traditional gender stereotypes that lead to multiple issues of gender inequality, objectification, and marginalization. As tattoos and the tattoo culture continue to enjoy popularity and become more mainstream, the issues of how gender is presented will remain important. The proliferation of digital space sharing and self-representation platforms provide the opportunity to continue and expand research on how social media and cooptation affects the tattoo culture at large.

References

Adams, Josh. 2009. "Marked Differences: Tattooing and its Association with Deviance in the United States." *Deviant Behavior* 40:266–292.

Agger, Ben. 2012. *Oversharing: Presentations of Self in the Internet Age*. New York: Routledge.

Atkinson, Michael. 2002. "Pretty in Ink: Conformity, Resistance, and Negotiation in Women's Tattooing." *Sex Roles* 47(5/6):219–235.

Barron, Lee. 2017. *Social Theory in Popular Culture*. London: Rowman and Littlefield.

Barron, Lee. 2013. *Tattoo Culture: Theory and Contemporary Contexts*. Basingstoke: Palgrave Macmillan.

Bourdieu, Pierre. 1984. *Distinction: A Social Critique of the Judgment of Taste*. Cambridge, MA: Harvard University Press.

Braunberger, Christine. 2000. "Revolting Bodies: The Monster Beauty of Tattooed Women." *National Women's Studies Association Journal* 12(2):1–23.

Burns, Deborah Louise. 2014. "Presentation of Gender." MA Thesis, Department of Sociology, California State University Los Angeles.

Baym, Nancy K. 2010. *Personal Connections in the Digital Age*. Cambridge, MA: Polity.

DeMello, Margo. 2000. *Bodies of Inscription: A Cultural History of the Modern Tattoo Community*. Durham: Duke University Press.

Fisher, Jill. 2002. "Tattooing the Body, Marking Culture." *Body and Society* 8(4):91–107.

Goffman, Erving. 1963. *Stigma: Notes on the Management of Spoiled Identity*. New York: Simon and Schuster, Inc.

Hatton, Erin and Mary Nell Trautner. 2011. "Equal Opportunity Objectification? The Sexualization of Men and Women on the Cover of Rolling Stone." *Sexuality and Culture* 15(3):256–278.

Hebdige, Dick. 1979. *Subculture: The Meaning of Style*. London: Routledge.

Kosut, Mary. 2006a. "An Ironic Fad: The Commodification and Consumption of Tattoos." *The Journal of Popular Culture* 39(6):1035–1048.

Kosut, Mary. 2006b. "Mad Artist and Tattooed Perverts: Deviant Discourse and the Social Construct of Cultural Categories." *Deviant Behavior* 27(1):73–95.

Mifflin, Margot. 2013. *Bodies of Subversion: A Secret History of Women and Tattoo*, 3rd edition. New York: powerHouse Books.

Pew Research Center. 2010. *Millennials: A Portrait of Generation Next.* Retrieved November 19, 2017 (http://assets.pewresearch.org/wp-content/uploads/sites/3/2010/10/millennials-confident-connected-open-to-change.pdf).

Sanders, Clinton. 2008. *Customizing the Body: The Art and Culture of Tattooing.* Philadelphia: Temple University Press.

Santos, Xuan. 2009. "The Chicana Canvas: Doing Class, Gender, Race and Sexuality Through Tattooing in East Los Angeles." *Feminist Formations* 21(3):91–120.

Thompson, Beverly Yuen. 2015. *Covered in Ink: Tattoos, Women, and the Politics of the Body.* New York: New York University Press.

Trottier, Daniel. 2014. *Identity Problems in the Facebook Era.* New York: Routledge.

Viren, Swami and Adrian Furnham. 2007. "Unattractive, Promiscuous and Heavy Drinkers: Perceptions of Women with Tattoos." *Body Image* 4(4):343–352.

Zones, Jane S. 2005. "Beauty Myths and Realities and their Impact on Women's Health." Pp. 65–80 in *Gender Through the Prism of Difference*, 3rd edition, edited by M. Baca Zinn, P. Hondagneu-Sotelo and M.A. Messner. New York: Oxford University Press.

3

POWERING DOWN

Theoretical Lenses to Examine the Agency of Our Smartphones

Daniel Okamura

Relative Power and Social Agency

Smartphone usage has increasingly come under scrutiny, often highly critical, as smartphone ownership and engagement among millennial adults and emerging adults has rapidly increased. In fact, millennial smartphone ownership is at near saturation levels, as "more than nine in-ten Millennials (92%) own smartphones" (Jiang 2018). Further, a significant number (39%) describe themselves as being "almost constantly online" (Perrin and Jiang 2018), while a quarter (25%) report spending more than five hours a day online (Globe Newswire 2017). The current research started from an interest in how the prevalence of smartphones affects the population of owners with the highest degree of adoption—the millennial generation. However, the author does not intend to add to the moral panic narrative around smartphone dependency. Rather, the purpose is to explore four theoretical approaches that may offer insightful ways of understanding the relationship between smartphone users and their phones. The analysis is not based on the assumption that smartphones hold ultimate power over people or that Millennials are slaves to Samsung, Apple, or Google. Rather, the hourly journal entries of Millennials who abstained from using their smartphones (and any other digital access devices for twelve consecutive waking hours) were used to indicate which theoretical approaches might be most fruitful to explore.

Smartphone ownership changes individual behavior in observable ways. For instance, almost a third (31%) of those who access the Internet from a mobile device say that they are online "almost constantly," whereas "among Americans who go online but *not* via a mobile device … just 5% say they go online almost constantly" (Perrin and Jiang 2018). Further, more than a quarter (28%) of Millennials are "smartphone only Internet users" (Jiang 2018). At a more concrete level, the current research conducted with

Millennials giving up their smartphones for twelve hours showed, for example, that some forgot their most frequently used phone numbers (9%) or how to get to familiar destinations by car (6%), but this form of dependence is not the same as being controlled. In other words, they recall fewer phone numbers from rote memory, but they determine what kinds of information they want stored in their digital address books. They may grow accustomed to having turn-by-turn GPS navigation, but they influence which directions and maps are the most accurate. All of these statements are true to some extent, which means researchers can, and must, use varied approaches to framing data interpretation. In so doing, some contradictions may arise—dynamics that ring true in some circumstances but not others. Each lens presented answers one central question differently: between people and technology, who has the power? By shifting the power dynamic, different truths emerge in the data.

First, *agency* is defined as *the ability to act independently*. Most people think of it like free will—the ability to make one's own decisions—but agency also includes the ability to act upon those decisions. It is often contrasted with *structure*, which is usually defined *as environmental factors that limit an individual's choices*. When social agency is contrasted with social structure, the central issue concerns patterns of group behavior in a given society that limit individual choices and actions. However, this chapter, focuses on the degree of agency all people within a given group have, as a collective, and uses the word *agency* interchangeably with the word *power*.

Technological Determinism

Philosophical theories that assume that human beings have no agency are grouped together as *determinism*. This means that all events are determined by pre-existing causes, as in the laws of cause-and-effect—what goes up must come down due to gravitational pull. When people are added into the mix, however, determinism becomes awfully murky. After all, people make their own decisions—or at least they think they do. This is where social thinkers have tried to assert that parts of the social world have deterministic effects. Biological determinists have asserted that our composition as animals determines what happens in our lives. They might argue that taller people, for example, would be better basketball players. Cultural determinists have argued that cultural socialization determines the way people think. Male friends in many countries may walk down the street holding hands and wonder why those in Europe and the United States hardly ever touch. Upon first glance, these examples appear to make sense, yet they have repeatedly been observed to be untrue. The National Basketball Association has had many successful players as short as five-foot-three. Travelers and migrants become accustomed to norms and behaviors of different societies everyday. Most social scientists agree that theories of determinism do not adequately explain variations in human thought and behavior. However, scholars in the interdisciplinary field of science and technology studies (STS)—and even some sociologists—recognize "technological determinism" (TD) as a useful boundary in examining the development of the world.

Some *technological determinists* argue that technology develops independently from human activity, and that it is the primary driver of what happens in human history. Under that assumption, others take a less direct stance on what drives human history and merely claim that technology can provide a baseline or an upper limit to our capabilities. Technological determinists can agree, however, that technology influences a society's social structures, cultures, and values (Paragas and Lin 2016).

Data from the Millennials who gave up digital access reveal phenomena that can both directly support and directly contradict a TD perspective. Relationships were tested by calculating Pearson's correlation coefficients. Participants who gave up their mobile phones and tried to use landlines to make phone calls (19%) could not recall people's phone numbers, because smartphone "contacts" lists are organized by name, not by phone number. Those who commented on giving up their phones' GPS navigation capabilities found they could not remember directions—even to places they regularly visit. One participant even boldly expressed a technologically deterministic view in the journal entry by asserting that "These inventions are near impossible to live without because societies' changes go hand-in-hand with technology." In fact, many participants described their experiences as "caused" by their phones. Their language reflects a sense of lost agency in view of smartphone technology:

> Technology seems to take control of all of our senses and I think I really noticed.
>
> As the day progressed I figured out that technology was holding me back from the actual things that I ought to do instead.
>
> Technology can make a person forget what it's like to just go out there and live your life.
>
> I had no idea that my phone was so close to me and that it has so much power.
>
> Digital media has become a mainstream way of communication in our society and in doing so can isolate anyone not onboard with the different social media outlets.
>
> I didn't think the phone was controlling my life but it is .... I felt a little scared because I finally realize how much my phone has interfered with my life. I did not think I would let my phone do this to me.

It is important to remember, however, that the most common capabilities of smartphones—email, text messaging, browsing the Web, playing music and video games, GPS navigation, taking photographs, and electronic calendars and appointment books—all previously existed. Not only did they exist in non-smartphones (so-called "feature phones"), but they existed outside of telephony altogether. This certainly suggests that TD cannot fully explain the patterns in this data.

Social Construction of Technology

Shifting from giving technology all the power to giving people all the power, different patterns emerge in the data. One family of theories argues that all reality is socially constructed. This means that nothing has any inherent meaning until people give it some kind of meaning. In this *social construction of reality* approach, objects and behaviors do not have any meaning until someone teaches us what those meanings are. Over time, groups of people agree on those mutual understandings. This group of ideas helps to explain why different societies develop different cultures. What may be a hand gesture of approval in one society may be a rude and insulting gesture in another.

Applying this social constructionist perspective to technology, yields the *social construction of technology* (SCOT). Advocates of SCOT argue that people and society control the design and uses of technology. Turning that statement around, one could also argue that technology is used to fulfill social functions (Woodhouse 2005). Some participants expressed this view fairly directly:

> The technology has shaped our line of communication with others, giving us easier access to either tweet or text someone at our convenience. The biggest significance of the millennial age is that technology has changed the way we choose to interact with others and the lifestyle we choose to live.
>
> Communication has improved; more and more people are texting, emailing, video-chats, etc .... Determining whether it's good or bad is up to those who use their devices.

This theory appears to explain a very large pattern in our data. Participants whose actions illustrated this perspective tended to seek out analog alternatives that served similar functions to the digital technology they forewent. Those who gave up video streaming sites like YouTube and Netflix replaced it with network television (77%). Those who gave up digital music, including streaming and mp3 files, listened to the radio instead (47%). Since the participants were students, they read books instead of using electronic readers (49%). They also paid cash for things instead of using electronic payments (49%), though many of them forgot to draw out cash before beginning the study. Generally, however, they found such analogs lacking, often speaking in terms that reveal the importance of controlling their environments, for example, one participant complains:

> Rather than accessing the Internet on my phone or computer for instantaneous access to around the world news in various venues, I was forced to turn on the TV to channel 11. I felt limited and extremely annoyed … Rather than providing information that I felt was relevant, I was forced to listen to information I was not interested in.

Nearly one in five (19%) mention using their phones as a means of "tuning out" unwanted stimulation, distractions, or interactions. One participant mournfully echoed a popular notion that smartphones are isolating us from one another. According to this person, "At home, when all my sisters are using their phones, we don't even talk to each other." While this idea of a phone as a kind of shield resonates with a lot of people, the behavior is hardly the device's fault. People have used newspapers, books, magazines, other people, and literally anything within their immediate surroundings to block others from interacting with them for hundreds of years. Think of a pet dog or a toddler: they may do nothing more than look away and pretend to not have heard you to keep from interacting with you. This behavior is not new.

This analytical frame, like TD, is also not perfect, and opens up new questions. What social function does checking a social media site serve? The data only showed a correlation between giving up social networks (including Facebook) with photo sharing (including Instagram). In fact, there were no correlations among participants between giving up their smartphone, sending each other text messages, or using social networks, and actually talking to people. This could suggest that people did not fully consider texting and social networks to be social or communicative acts. Another possible SCOT interpretation can be that what is given up is not analogous to conversation but to outward announcements and other types of messages that enable the sender/poster to control the narrative. Additionally, it can be that mediated messages enable a lag in time not available in conversation that enables those messaging to gain more control over framing a response than they have in the immediacy of conversation. As one participant put it:

> On a daily basis I only make small talk with people instead of sitting to connect with someone. When I did start having conversation with people, I didn't even know how to act … [I felt] the inability to send messages across and establish understanding.

Similarly, it has been observed that texting, for example, has gained popularity in part because "people want to customize their lives, they want to go in and out of all the places they are because the thing that matters most is control over where they put their attention" (Turkle 2012). Participants reflect this in their journal entries.

> I text and am on my phone at all times no matter where I am …. My full attention is never all given. I wanted to be connected to all different places …. I do not make any personal appearances, I just text. I felt like I wanted to be close to people but only by my social networks …. I'd rather text than talk.

This recasts phone-based messages as quite distinct, rather than being a parallel to actual conversation, and ties back to individual power over the uses of technology.

There are also social functions that are not directly connected that we may be missing. One participant in the study noted that their morning routine involved taking and posting a photo of their dog on Instagram. When they did not post a photo that day, their friends grew worried and tried to contact them. This suggests that the mere act of posting on social media, regardless of content, could be seen as a kind of beacon or ping that lets others know you are all right—or even that you exist altogether. Another participant noted:

> My friends were upset because I was not responding to their texts or calls even though their calls were not something of great importance. Due to my quick text responses before, they thought something was wrong.

TD is often depicted as the polar opposite of SCOT. However, rather than view this as an either/or proposition, it has traditionally been more common and more fruitful to consider these as endpoints of a continuum, with technology controlling individuals at one end and individuals exerting control through technology at the other end (Paragas and Lin 2016). Additional theoretical approaches fall in between these two extremes, and some scholars argue they do not negate the usefulness of these perspectives. The need for other approaches is based, in part, on the fact that agency occurs in a dynamic context. In other words, although individuals may make decisions and take decisive actions, they cannot always control the outcomes of their decisions and actions. Looking at it this way, their power is limited, but not absent. Ultimately they may have only imperfect agency.

Actor–Network Theory

The results of an individuals' actions are not always directly related to their intentions, which can mean that people are not fully in control of the way events unfold—that even the most decisive individuals may not always have complete power in a given situation. Such a scenario requires a theoretical approach that can explain reality considering that people are not fully in control. This has historically made theologians and philosophers uncomfortable, as those fields of study tend to believe that free will must be complete and unfettered in order to be free will at all. Adherents of more controversial, contemporary schools of thought, may argue that just because people are not always in full control of what happens to them and because of them, does not mean that they lack all agency. One such theory is called *actor–network theory* and suggests that nonhuman creatures and inanimate objects can be partially agent as well. This is not to say that a door or a glass of beer are capable of making their own decisions. Such nonhuman or inanimate *actants* are able to restrict people's behaviors and even act back upon them (Jerolmack and Tavory 2014). Think of when someone expecting to go through a door finds it locked instead, or when alcohol has gotten a person more intoxicated than they had expected. The bee stings the person tending to a colorful garden. A shark eviscerates the vacationing surfer. In these contexts, these actants have affected the outcome of an individual's actions and, subsequently, change future interactions in ways the individual did not intend.

A small number of participants in the study discussed their phones in ways that depicted smartphones as actants. These participants did not just reify their phones as something they interacted with, but even began to anthropomorphize them as something to which they felt emotional attachment. Journal entries often expressed general longing: "As I got ready for bed, I began to really start to miss my phone." Others, however, found their phone much more of a central part of their life, orienting their thoughts around it and expressing anxiety. For example, "I'm just counting the hours and minutes until I get my baby back and, by baby, I mean my phone. I need it." Similarly, another participant recognized that she missed the device more than the person she most wanted to contact, "I am starting to think that my anxiety might have more to do with the separation of my phone, rather than my boyfriend." In contrast, another participant's journal entry expressed that digital abstinence brought this type of personification to her attention as undesirable:

> I would find myself with my phone in my hand, even if I was not using it. I felt it was necessary for me to know that my phone was in sight at all times. Feeling this way is not right. Why do I need to hold my phone? It is not a baby that needs to be looked after.

One of the strengths of actor–network theory is its ability to explain people's relationships with nonhuman creatures and objects on a *micro-social* level, meaning on the level of individual and small group interactions. Though it has been used to describe how groups of people orient themselves around objects and ideas, it has not widely been used at a *macro-social* level, in order to explain large moving organizational parts and structures in a given society or between societies.

Social Shaping of Technology

For a *macro-social* explanation, the theory of *social shaping of technology* (SST) may help. Proponents of SST argue that society and technology are not separate entities. Instead, they are part of the same trajectory of history—and a nonlinear one at that—where social conditions and technological advances inform one another in a kind of *mutual shaping*. SST has been used to explain the relationship between technology and society in a cyclical manner. Some participants recognize this mutual shaping with acknowledgements such as this journal entry:

> I have a much greater appreciation for all the technological advances that I am so fortunate to have in my life. Having been able to enjoy much of my adolescence with a lot of toys, video games and computers available, I would be like a fish out of water if it were all to disappear. Everything about my life has some sort of technology involved. The real lesson learned in this assignment was the gratitude towards things I would have taken for granted—my phone, my computer, directions that are used in my navigation system—[that] have helped things to be more convenient.

SST examines the effects of a particular technology on a society but also examines the conditions that make certain technological advances possible (Boczkowski 2004). Social conditions influence what kind of technology develops and what form it takes. Once technological advances are made available for research, industry, consumer, or government purposes, it affects how those areas of society develop. In turn, certain features in technology become popular, while demand for still other features affects how future technology develops. On the other hand, some features fail to catch on, and that stifles their future development. For example, the first software mp3 players were released in 1997. Until then, Americans had no use for digital music files. However, the availability and the popularity of the software created conditions for technology companies to exploit, which led to the development of portable mp3 players. One such player became very popular because of its slick aesthetics based on mid-century modern design. When this device, the iPod, was released in 2001, demand for digital music files skyrocketed. This shows technology having a significant effect on society. After such increased demand for digital music, people began to choose digital files over compact discs to the point where media companies make more money from streaming subscriptions than from CD or download sales (see Chapter 10). This shows social conditions affecting what kinds of technology get developed, what becomes popular, and what fails. This example shows SST's ability to explain the cyclical and iterative nature of society and technology shaping one another.

Since the current study essentially captured one specific period of data collection, it is not particularly suited to theories like SST. However, researchers may look at it as one point in time, and later examine how both the technology and society change after that point. The participants were nearly unanimous in remarking on having to give up their smartphones—96 percent, as compared to other digital access devices. Further, almost two-thirds (60%) described extremely strong cravings for their phones—almost seven times the number who craved digitized music (9%) or computer use (9%). This is explained by the fact that for most, music and Internet access are more frequently part of phone use, as compared to mp3 players and computers. In other words, those speaking in terms of what they craved most are reacting to the multiple uses of their phones, for example:

> I rely on it for virtually everything, from directions to getting my daily news—I do find it important to be up to date on what goes on in the world … I not only depend solely on my phone to gain access to the news and my friends, but also to music, something I enjoy quite frequently.

While nearly everyone in the study remarked on the absence of their phones, the activities they used their cell phones for were far from uniform. Most participants (80%) mentioned that they missed listening to music on their devices. This is a high frequency, but nowhere near the universal number of those who missed their phones, overall. Only two-thirds (69%) of respondents commented on the absence of social networking—odd, as the global discourse around social networking treats it as pervasive and everywhere. More than three-quarters (79%) of participants commented on the inability to send text messages and email and more than half (55%) said they missed watching YouTube videos. This may, in part, be explained by the idea that many may state that they miss their smartphones without explaining which of its many functions they miss.

Because SST examines changes over time, it is helpful to remember that these are all uses of mobile phones that existed before the smartphone. Even if these numbers are surprisingly low, these features were probably popular, and smartphone and software developers sought to keep them or improve upon them as the shape of mobile phone technology changed. SST may prove a useful tool to analyze whether these numbers change, whether phone capabilities homogenize, or whether new uses emerge.

SST may also be useful in examining how technologies become expected or transparent in a given society. This is one of the reasons the millennial generation is particularly important to study: the tail end of Generation X remembers what life was like before widespread Internet use, whereas Millennials include the first true digital natives. Facts existed before the ability to "Google it," but much of the developed (and developing) world has arrived at a point in time where people expect others to have smartphones. Most phones did not have turn-by-turn navigation software at the beginning, either—that did not exist until 2010 for Android users and 2012 for iPhone users. One respondent from 2014 recounted having to stop and ask someone for directions in the absence of his phone—the person they approached asked, "Why don't you just use your phone?" It certainly did not take long for turn-by-turn navigation to become transparent in American society.

Conclusion

Four theories with which to analyze the data in the offline millennial journal study have been presented: TD, SCOT, actor-network theory, and SST. Admittedly, one of them, TD, is merely rhetorical, but its philosophical foundations inform the other three. By shifting the focus of who has more agency within a given situation, different questions may be asked of the data and different patterns of behavior come into relief. Such a multifaceted approach helps to reveal the complexity of social phenomena and avoid reductionist explanations. Of course, other theories exist. For example, normalization process theory focuses on practices, and may help illustrate how people use technology in particular social settings. In social psychology, the *locus of control* (LOC) concept argues that it is the individual's personality that explains whether technology is seen as an external force that controls individuals, or whether individuals believe that they control the impact that technology has on them. LOC, alone, however, is unable to explain trends and common group behaviors or broad societal outcomes (Durak 2018; Shahibi and Aziz 2017). There is much room for further research on how smartphones and other pieces of technology affect social and individual behavior, whether this involves using the theoretical frames described here or by examining and following the data longitudinally.

References

Boczkowski, Pablo J. 2004. "The Mutual Shaping of Technology and Society in Videotex Newspapers: Beyond the Diffusion and Social Shaping Perspectives." *The Information Society* 20(4):255–267.

Durak, Hatice Yadiz. 2018. "What Would You Do Without Your Smartphone? Adolescents' Social Media Usage, Locus of Control, and Loneliness as a Predictor of Nomophobia." Addicta, February 27. Retrieved June 26, 2018 (http://addicta.com.tr/en/article/what-would-you-do-without-your-smartphone-adolescents-social-media-usage-locus-of-control-and-loneliness-as-a-predictor-of-nomophobia/).

Globe Newswire. 2017. "Smartphone Obsession Grows with 25% of Millennials Spending More than Five Hours Per Day on the Phone." May 18. Retrieved

June 26, 2018 (https://globenewswire.com/news-release/2017/05/18/987607/0/en/Smartphone-Obsession-Grows-with-25-of-Millennials-Spending-More-Than-5-Hours-Per-Day-on-the-Phone.html).

Jerolmack, Colin, and Iddo Tavory. 2014. "Molds and Totems: Nonhumans and the Constitution of the Social Self." *Sociological Theory* 32(1):64–77.

Jiang, JingJing. 2018. "Millennials Stand Out for Their Technology Use, But Older Generations Also Embrace Digital Life." Pew Research Center, May 2. Retrieved June 22, 2018 (http://www.pewresearch.org/fact-tank/2018/05/02/Millennials-stand-out-for-their-technology-use-but-older-generations-also-embrace-digital-life/).

Paragas, Fernando de la Cruz, and Trisha T.C. Lin. 2016. "Organizing and Reframing Technological Determinism." *New Media and Society* 18(8):1528–1546.

Perrin, Andrew and JingJing Jiang. 2018. "About a Quarter of US Adults Say They Are 'Almost Constantly' Online." Pew Research Center, March 14. Retrieved June 22, 2018 (http://www.pewresearch.org/fact-tank/2018/03/14/about-a-quarter-of-americans-report-going-online-almost-constantly/).

Shahibi, Mohd Sazili and Fazlina Abdul Aziz. 2017. "The Effects of Smartphone that Influence the Compulsive Usage Among Students." *International Journal of Academic Research in Business and Social Sciences* 7(8). September. Retrieved June 26, 2018 (https://www.researchgate.net/publication/320224830_The_Effect_of_Smartphone_that_Influence_the_Compulsive_Usage_among_Students).

Turkle, Sherry. 2012. "Connected But Alone?" Ted Talks, April 3. Retrieved June 26, 2018 (https://www.youtube.com/watch?v=t7Xr3AsBEK4).

Woodhouse, E.J. 2005. "(Re)Constructing Technological Society by Taking Social Construction Even More Seriously." *Social Epistemology* 19(2–3):199–233.

4

FROM BACKSTAGE TO DIGITAL FRONT STAGE

Online Queer Community, Identity, and Emotion Management

Lacey M. Sloan

Relative Marginalization and the Queer Community

A new wave of human rights has begun to focus on a particular group of people who have largely been marginalized on the basis of gender identity and sexual orientation. The branding of these persons as deviant is unevenly applied, as it depends on an individual's location within the queer community. The discussion is ever changing, with revolutionaries providing perspectives on the micro-social strategies of the marginalized, and on the macro-social process of shifting norms. *Micro-social* issues of identity formation, self-presentation, mental health, and virtual community are examined against the backdrop of cohort impact on progressive change. Gaining knowledge about those considered to be outside of the social norm is crucial for understanding social order, disorder, and change. For the purposes of this study, the context that marginalizes others is the *heteronormative* view that dominates mainstream beliefs, values, and perceptions in which social expectations are predominantly built around the *assumption that heterosexuality is the "normal" orientation*. Understanding this context is key to fully understanding social norms and the pressure to comply. Further, neither social compliance nor social deviance can be understood independently of one another—one needs to know what it means to comply in order to recognize what constitutes deviance.

Social compliance in a given community, culture, or society, must be studied in view of established rules, expectations, and guidelines. This chapter looks at how these rules, expectations, and guidelines are manifested in communication, relationships, and gender, within the context of rigidity, rejection, and backlash. These rules are maintained, enforced, or negotiated in constant reference to a narrative about the greater good and what is considered detrimental to the status quo. Social deviance then exists because these rules and guidelines will inevitably be broken. In the *heteronormative* context, this means that anyone along the queer spectrum is considered deviant. Social compliance and deviance, however, are in constant motion, being formed by and against one another as newer generations change their ideas, expectations, and beliefs.

One of the most recent and significant developments can be seen as Millennials increasingly accept, and show compassion toward, individuals belonging to the LGBTQ+ community. The *LGBTQ+* community is comprised of individuals who identify as lesbian, gay, bisexual, transgender, and *queer* (in this chapter, queer will be used as an umbrella term to encompass questioning, asexual, non-binary, intersexual, undecided, genderqueer, pansexual, and everything in between). The LGBTQ+ label has become a familiar identifier (at least among academics); however, the most normative shifts among non-Millennials have been concentrated on the LG portion of that larger community (Allen 2015). Despite Generation X and Boomers increasing support for same-sex relationships (Pew Research Center 2015), the more expanded concepts of gender identity still tend to be considered taboo.

This progress, however, is slow moving and selective, forcing most members of the queer community to deal with a conservative backlash against physical and mental health services, media and historical erasure, and other means of discrimination related to an individual's location on the queer spectrum. Unfortunately, while the 2015 Supreme Court decision for marriage equality was an example of a step in the right direction, other much-needed resources are scarce and face-to-face safe spaces are essentially nonexistent, making the Internet a crucial resource for queer youth. Online blogging sites and queer-centric groups and accounts, from Facebook, Tumblr, Instagram, and Twitter, have quickly become places for fostering identity formation, self-presentation, mental health, and community.

The data collection for this study focused on the importance of identity formation, self-presentation, mental health, and a sense of community within these online asylums for members of the LGBTQ+ population. The findings and the implications of the results are considered within the larger social-psychological frameworks by providing support for previous research and the idea that members of the queer community would find online comfort with both themselves and others. However, the use of humor as a coping mechanism in each of the four themes was far greater than anticipated, and division and toxicity within the community was both prevalent and unexpected.

Findings and Discussion

This study was based on a content analysis of 406 text posts, photos, images, videos clips and memes collected from openly queer-identifying blogs, handles, and public pages through Facebook, Twitter, Instagram, and

Tumblr. Such content was categorized into relevant themes and sub-topics. Findings are discussed in terms of how these themes and or sub-topics illustrate important considerations about queer identity formation, self-presentation, mental health, and community. Overall, the content analysis supported previous research regarding the prevalence of mental illness within the LGBTQ+ community, and social interaction within online communities. However, three unexpected inferences can be made from the thematic patterns that arose. First, self-acceptance and ownership of one's identity was surprisingly salient, as it somewhat contradicts the proven mental health crisis phenomenon among queer individuals. Second, humor, especially humor reliant on hyperbole or sarcasm, was present in over 90 percent of content. Lastly, some content pointed to problematic aspects of the community—meaning some users posted and re-blogged content about their need for a break from the community because of the potential toxicity.

Identity Formation

According to Berzonsky (2005), a coherent, stable sense of identity provides people with a sense of purpose and serves as a personal standpoint for interpreting experience, solving problems, and making life decisions. When such coherence and stability are connected to positive self-evaluation, identity formation can be seen as a lynchpin of making positive life decisions and solving problems in psychologically healthy and pro-social ways. Identity formation is not only a critical part of how individuals enact their daily lives, but also of how they impact their social environment and the lives of those around them. An example of this can be seen in the amount of content focused on questioning of the nature of the human condition.

Over 70 percent of the content that was relevant to identity formation dealt with the meaning or the meaninglessness of existence; these were grouped together due to their philosophical nature. Users questioned their existence, mainly by comparing themselves to the vastness of the universe, and how everything around them is socially constructed. For example, one meme was accompanied by an image of a person peacefully asleep in bed, floating among the clouds with text that read, "How I sleep at night knowing none of this exists, nobody belongs to me & people's actions are merely projections of their own issues." Another stated, "We're literally floating on a tiny planet in fucking space why are we surrounded by hatred and misery …?"

Much of the content emphasized the idea that individuals have control over how they perceive their reality, whether good or bad (Burnham and Papandreopolous nd). On the other hand, the content also expressed the philosophical concept that life itself is meaningless, and that nothing in the world has any real existence. This ability to see oneself within the wider perspective of human existence—especially through a social construction of reality lens—can act as a grounding factor. Ideally, such grounding can release the individual from the constraints of societal standards and expectations, and enable the letting go of anger, disappointment, and anxiety.

Unfortunately, such release is not a simple matter. Miller (2016) explains that members of the queer community live with non-visible stigmatized identities, and attempting to manage such identities can be isolating due to the secretiveness involved. Devastating outcomes can include micro-social level experiences (such as stigma; rejection by friends, coworkers or family; and the fear of potential rejection); and macro-social level factors (like the awareness of, or personal experience with, gender-identity based discrimination). Meyer (2003) explained that conditions in the social environment, as well as personal events, are sources of stress that may lead to mental and physical ill effects. Social stress has a strong impact in the lives of people belonging to stigmatized social categories, in this case, the queer community.

The perceived need for secrecy can lead to a battle among multiple identities, especially when one feels compromised. While little content clearly fit into the self-rejection category, it is implicit in some imagery. For example, an image of a puppet, eyes wide with shock states, "When someone calls you by your birth name," followed by "What? Who's that?" The birth name represents both an identity that a transgender individual no longer identifies with, and it indicates someone they once were. An individual may be attempting to suppress or reject an old identity to create a new, better-fitting one. This new identity, however, may not always be confirmed by others, and could create a sense of invalidation, dysphoria, and rejection.

Another factor that contributes to the formation of identity are feelings and emotions. Feelings and emotions are a response to stimuli that create a positive-to-negative evaluation of one's state of being (Gable and Harmon-Jones 2013), including the evaluation of self. However, it is clear that a multitude of life events shape ones identity, and that positive and negative stimuli come from all directions. Data show that individuals whose feelings and emotions have been triggered by negative social judgment of their gender identities, can avoid negative evaluation of self by changing to more accepting virtual communities that offer positive stimuli. In fact, nearly all of the content associated with positive identity formation dealt with having one's identity align with others' perceptions. For example, a meme of He-Man in front of a rainbow backdrop, throwing his head back and laughing, framed with the text, "tfw [that feel when] people gender you correctly and use your preferred pronouns." Positive stimuli from more accepting communities would then, presumably, have a positive influence on feelings of happiness and contentment in association with identity formation.

An evaluation of previous research would imply that the content circulated online through the community in regard to identity formation would be negative due to the stigma, discrimination, and isolation that goes along with the queer experience. However, a significant subset (18%) of the identity-formation content was associated with a positive sense of self, or self-acceptance. Content ranged from letting go of negative self-thought, through accepting oneself regardless of others' perceptions, to exaggerated compliments about how gay and fabulous one was. It seems as though individuals share online content to build more positive senses of self, despite the many sources of negative stimuli.

Meaningful lives are judged to be valuable, worthwhile, understandable, and a means for making an impact in the world (Negru-Subtirica et al. 2016:1927). Further, identity construction takes place as individuals commit to specific life goals and values, while attempting to see their lives as meaningful. Without individual psychological health as a positive basis for decision-making, however, establishing what might be deemed "a meaningful life" becomes problematic. Marginalized populations find that developing coherent, stable, and positive-self-evaluation-based identities is challenged by discordant social expectations and definitions of deviance in face-to-face interactions. This leads many to seek an online niche for support in creating a meaningful life.

Contrary to previous literature on the queer experience, the self-acceptance content in the identity formation data implies that members of the queer community see their lives as meaningful. At the very least, data indicate that they have found a means of dealing with the issues that accompany queer identity. However, identity formation cannot be examined without also examining interactions and self-presentation. After all, interactions with others help dictate just how visible one's identity is.

Self-Presentation

In order to navigate the social world, individuals engage in *self-presentation* to express desired identity images to particular audiences. Erving Goffman, a pioneer in the analysis of social interaction as self-presentation, defined it as a balancing act between highlighting positive aspects and reducing negative impressions in an attempt to constantly be seen in a positive manner (1956:14). This type of impression-management varies across social locations such as households, schools, churches, and work settings; generally, however, there are also great commonalities that result in a relatively consistent presentation of self.

Study results are consistent with the literature, showing that individuals made sense of their identities through careful self-presentation, which may vary in the degree to which it includes more authentic *self-disclosure*—that is, the sharing of meaningful (often private), defining characteristics, beliefs, and experiences (Michikyan, Subrahmanyam, and Dennis 2015:288). Twenty-six percent of the content in the self-presentation category dealt with external validation through interaction with others. Two particularly strong illustrations of these approaches follow:

> Finally being ok with the knowledge that u don't actually have a personality, just different facades u put on for the different people in ur life, but now u cant stop thinking about the facades ppl put on around u and whether or not anyone genuinely knows another person or if we are all just empty shells playing pretend.

The other image is of a pair of high heels adorned with keyboard keys across the straps accompanied with the text:

> When you realize Internet identity is performative, but you have also made many deep, lasting friendships and met most of your intimate partners online so you begin to wonder if you're actually performing your truest self.

These show the complicated reciprocating relationship between identity formation and self-disclosure. Internet friendships and relationships often form because offline circumstances, creating deep understanding of one's own self-presentation and identity, alongside others' self-presentation and identity.

Maintaining a favorable identity is a common goal for most individuals; however, the ability to control and manipulate others' perception can be crucial and potentially life-saving for marginalized LGBTQ+ persons. For members of the queer community who rely on an awareness of self-presentation, an appropriate public image could mean hiding their true gender identity or sexual orientation and presenting a more heteronormative, or *passing* identity. A passing identity relies on relevant social scripts to express behaviors that are more context appropriate or normative (Tyler, Kearns, and McIntyre 2016:175). The ability to present a straight, heterosexually passing identity is more than just the ability to dress in a traditionally gendered way; it also requires one to act and speak in mannerisms that fit within the expectations of the gender dichotomy. It is the preferential power given to this gender dichotomy as normative that defines the mainstream culture as *heterocentric.*

This can be seen in the prevalence of content relating to gender performance, which made up well over half (60%) of the self-presentation content category. An image of singer Mariah Carey, young and smiling, framed by the words:

> Preparing for a long day of feminine performance by applying a heavy mask of makeup to appease baseline beauty standards.

This illustrates the emotional expenditure needed to constantly uphold an acceptable feminine standard of beauty. Mariah Carey has long been a popular gay icon, frequently associated with glitter, dresses, and dramatized, hyper-feminine measurements, that illustrates how wearing makeup and performing to the standard of traditional femininity fits within the idea of the gender dichotomy.

The internal struggle to know and define oneself exists in constant relation to others, and that definition of oneself is constantly being performed, defined, negotiated, and maintained. Specifically, the queer identity is constantly being negotiated under heteronormative standards. One item of analysis, in particular, featured an image of Jane Bowles (a writer who has been celebrated for her feminist and lesbian literature) alongside her platonic husband, Paul, with text that read:

> When everyone you've ever loved has tried to change you, so you change the way you view partnership to exclude heterocentric standards of mirroring.

Balancing one's identity cannot exist without an external push–pull relationship with others. Personality is shaped by shared experiences with others, whether or not the experiences are positive.

However, negative experiences can push the personality into hiding and prevent queer individuals from fully expressing their sexual orientation or gender identity. One such experience cones in the form of *gender policing*—meaning the imposition of normative gender expectations on those considered not to be performing gender appropriately. Being closeted, or not having peers and loved ones know about ones sexual orientation or gender identity, can result in a disconnect from others. Individuals may suffer isolation and frustration if they are not able to express themselves, and be honest about who they are around others due to fear of judgment or rejection. For example, an image from the children's show *Arthur* featured an image of Arthur's tightly closed fist (which is frequently made into memes to express frustration) captioned:

> when you're closeted and your family members start to gossip & talk shit about lgbt acquaintances like some kind of straight bonding ritual.

This expresses that being closeted is still a very relevant problem for queer individuals, especially around family members.

Individuals who are expected to pass are often pressured to conform to especially rigid gender norms that would not necessarily be applicable to someone who is heterosexual and *cisgender* (identifying with the assigned gender at birth). Individuals might potentially restore threatened gender status by presenting themselves in ways that strengthen the perceived gender identity. Such presentations involve acting in socially acceptable ways for the gender dichotomy (Sinclair, Carlsson, and Bjorklund 2016:428). This could be especially problematic for transgender or other non-binary individuals who are expected to pass, as it puts restrictions on the options for self-expression and self-presentation.

However, the need to present a passing identity can be alleviated by the anonymity of the Internet. This does not necessarily mean being at ease, however. Once interaction moves from a physically-inhabited environment to a digital-space environment, the need for impression-management increases drastically. Social networking sites elevate the need for self-presentation management because the persons on such sites may come from various social settings that are no longer separated the way they are in the physical environment. A colleague may turn into a Facebook friend, or an Instagram follower may be a long-term, religious family friend (Duguay 2016:893). Duguay called this concept *context collapse* because social networking sites blur the location, time, and institutional boundaries that otherwise separate audiences (2016:892). This implies that not everyone has full control over when and who sees particular identity performances, which could result in someone having to come out to family members a lot sooner than intended.

This added pressure around the need for self-presentation management can increase the potential for problems, such as discrimination, familial rejection, or a disconnect from the community. More than a quarter (27%) of the content relating to self-presentation dealt head-on with the issues of having to protect oneself from societal gender policing. Two side-by-side images of actor Teraji P. Henson illustrate this. In the first image, she is dressed in expensive clothes, with her hair and makeup done well, with a menacing but pleased look on her face. The second image shows her hiding, crouched down holding the bumper of a car, looking fearful. The text at the top reads, "Being gay online vs. being gay in real life."

Oppositional images are also posted. One meme demonstrated a strategy for navigating traditional gender norms to one's advantage stating,

> When femme presentation goes from your default to a conscious, subversive act to fuck with the misogyny of both the straight and queer worlds you navigate daily.

This is shown alongside an image of a bikini-clad actor Mandy Moore. Here, femme presentation transcends gender performance, creating a sense of ownership and a reclaimed presentation. What little content was found regarding reclamation, embodied empowerment. In this case, gender expression is no longer considered a performance for others, but for the individual. Individuals learn to reject societal expectations, no matter how strong, and express themselves *for* themselves in a way that they find comfortable.

Mental Health

Mental health is defined as emotional, psychological, and social well-being that facilitates handling stress, relating to others, and making choices (MentalHealth.gov 2017). Considerable research has found a large mental health disparity among queer individuals on such issues as substance abuse, mood and anxiety disorders, and suicide (Bostwick et al. 2014:35). For those who are marginalized by their gender identities, this is complicated by extreme stress-inducing factors such as experiences with prejudice, stigma, the psychological battle between concealment and disclosure, and internalized homophobia (Meyer 2003:681–688).

Data support the literature; the majority of the content in the mental health category (61%) related to anxiety and depression. Population-based studies have shown that men and women reporting same-sex partners are more likely to use mental health services (DeAngelis 2002). Further, nearly half of transgender individuals report experiencing depression and anxiety (Schrieber 2016). Many examples from the data illustrate these issues; three examples follow: a text post that stated, "my mom might've raised an emotionally vacant child with severe depression and anxiety but she didn't raise a quitter;" an image of a person holding binoculars with text stating, "me looking for the serotonin my brain isn't producing;" and an image of an angry dog tucked into bed captioned, "waking up from ur depression nap to find ur still depressed." These examples depict queer mental health issues from the perspective of the queer individual.

Eleven percent of the content relating to mental health dealt with miscellaneous mental health issues such as borderline personality disorder and symptoms relating to trauma. It has been found that individuals with borderline personality disorder are significantly more likely to report homosexual or bisexual orientation and intimate same-sex relationships (Reich and Zanarini 2011). In addition, queer youth experience bullying, harassment, traumatic loss, intimate partner violence, and physical and sexual abuse, at higher rates than their straight peers (NCTSN 2015; Birkett, Newcombe, and Mustanski 2015). There is, however, little research regarding the proliferation of borderline symptoms in queer youth relating to such traumas. The content found regarding miscellaneous mental health issues dealt specifically with borderline personality disorder (bpd), trauma, or both. Content included "bpd bingo" with squares including symptoms that also overlap with symptoms linked to trauma, including self-loathing, self-harm, disassociating, sleeping for 20 hours, splitting, self-destructiveness, unstable identity, mood drop, and suicidal thoughts (Ozer et al. 2003). Noteworthy content, such as borderline personality bingo, can provide insight on how queer youths feel, regardless of official diagnosis.

LGBTQ+ individuals have experienced a fluidity in the categorization of mental illness in the past, starting with classification of homosexuality itself as a mental illness in the first and second editions of the Diagnostic and Statistical Manual of Mental Disorders (DSM). With each successive DSM, the diagnoses associated with queer individuals are shaped and defined by dominant social and cultural norms about sexuality and gender (Perone 2014:768–769). Some professionals continue to see any sort of deviance from heteronormativity as requiring treatment with aversive therapy, behavioral modification, or religious-based methods. Some others manifest a general lack of understanding of the queer experience, making it difficult for those who need therapy to pursue face-to-face treatment.

Outdated categories in mental health manuals or poorly constructed mental health campaigns, lead some queer individuals to resort to self-diagnoses, as indicated by one subtopic in the content (5%). While not always verified or confirmed by a professional, self-diagnoses can be incredibly helpful to personal growth and psychosocial development. Digital community members also offer possible diagnoses for one another based on disclosed information (Giles and Newbold 2011:424). Realization here is paired with validation, and the support of community members who emphasize the importance of identity for one another. For example, one text post simply stated,

> Self diagnosed mental illnesses are not less valid than clinically diagnosed mental illnesses!! you know more about yourself than other people do!! you're valid and so are your diagnoses!!

Self-diagnosis can dismiss doctors' professional opinions and lead to poor treatment choices regarding a certain illness. Many, however, have difficulty accessing mental health services and discussing their symptoms. For those with a sense of disenfranchisement from mental health professionals, beginning from the point of self-diagnoses may seem the only option. When individuals spend weeks, even months, perusing websites and other literature, they may find themselves more confident to pursue their needs.

Current societal perceptions of queer mental health are not without flaws. Queer youth mental illness and suicide is frequently treated as a weakness of the individual, and depicted as the person's inability to withstand the realities of the social world. Self-help discourse includes references to bootstrapping, competition, and entrepreneurship in exchange for a better life (Grzanka and Mann 2014:380). While the intentions may be good, testimonials and directives offered by certain mental health campaigns paint a picture of a life that may not be attainable to many queer individuals, including marriage, a family, stable housing, and a steady income. These approaches are meant to be encouraging, but often indirectly blame the individual for not attaining a social expectation for "happiness" and provide no real-life applicable advice for a positive, long-term trajectory.

One item, for example, contained a gif of a monkey puppet, looking shocked, with text that stated, "when a neurotypical suggests yoga and lavender and suddenly i'm Cured^{TM}." The sarcasm addresses the ignorance of simplistic approaches that cannot alleviate long-term, deep-seated, meta-health issues in the queer community.

Nearly as important as the diagnosis itself are the individual coping mechanisms comprising 16 percent of the content as a subtopic in the mental health category. These include, but are not limited to: distancing oneself from the stressor; planning stress-reduction strategies; limiting sources of distraction; developing self-restraint; waiting until the right moment to act; and disengagement through minimizing efforts to deal with the stressor (Carver, Scheier, and Weitraub 1989).

Several memes expressed actively choosing bad methods of coping, and engaging in self-destructive behavior. One item contained the image of a person licking a cactus, saying, "when u have a history of self destructive behavior and u know what ur doing is bad for u and not gonna end well but u do it anyway." Another text stated, "2017 we're all gonna stop posting self deprecating and 'I wanna die' memes." Someone else responded with, "you'll have to pry my morbid coping mechanisms from my cold, dead hands." Individual awareness of self-destructive coping mechanisms were accompanied by an undertone of humor. The ownership of maladaptive behavior could potentially be a response to current social perceptions regarding queer mental health.

Most of the content in the mental health category was focused on the negative aspects. Unexpectedly, however, a small portion of the content (7%) related to personal emotional growth in the mental health category of the content analysis. For example, one text post stated:

> Note to my ex. Today my professor told me every cell in our entire body is destroyed and replaced every seven years. How comforting it is to know that one day I will have body that you will have never touched … Important especially for victims of abuse, remember your body is yours and it heals in more ways than you realize. I reblog this every time I see it because every time, I'm a little more me and a little less you.

This type of content fits into what developmental psychologists broadly define as *resilience*, or a dynamic process wherein individuals display positive adaptation despite experiences of significant adversity or trauma (Luthar and Cicchetti 2000:858). Prior research shows that task-oriented coping and emotion-oriented coping are significant predictors of resilience (Campbell-Sills, Cohan, and Stein 2006:591–592). While this text post points to a physical aspect of resilience in particular, the ability to relate to the text is not without significance. Using comical content to address problematic behavior and reflect on resilience can potentially be used as a coping method for difficult experiences.

Community

The growing presence of young adults on social networking sites has fostered a digital environment rich with trust, communication, and self-growth. Digital-space communities provide positive interdependence as they strive for mutual goals and provide reference groups for individual participants (Bagozzi and Dholakia 2002:3). Some social networking sites allow for the separation of social groups, and can be seen as providing mental health and social support interventions. Though often unsophisticated and lacking in professional guidance, these online support groups offer alternative norms. Shared through personal anecdotes and advice, the online support can be divided along the lines of age, location, general interest or, in this case, queer identities.

Online queer communities proliferated due to the rejection of their members as social deviants in offline settings. The edging out of queer identities from heteronormative physical space, led to digital-space communities comprised of those perceived to be violating gender-relevant social norms. These digital communities can be essential for individual emotional functioning, as they address the conflicting needs for being able to affiliate with others while maintaining a distinctive, self-sufficient identity (Meyer 2003:681).

Individual identities come together through common interests, experiences, and most importantly, solidarity. Forty-two percent of the content in the community category referenced ones place within the community. One content item, for example, contained an image of widely known LGBT icons, singer Lady Gaga and television personality RuPaul, sitting side-by-side, hands folded as if to conduct an interview; the text stated, "When you renew your LGBT contract and have to face the board of directors." The message seems to imply that the queer community is exclusive, and one must perform their queerness to be a part of the club.

Members of this queer club validate one another's queerness by giving general "shout outs." General shout outs accounted for over a third (37%) of the content relating to community. Shout outs typically were text posts, with only a few containing images:

> Shout out to the gays! The who? THE GAYS!
>
> s/o [shout out] to kids that know they're probably not straight but won't let them self think about it. You'll get there.
>
> s/o to kids that know they're probably not cis but won't let them self think about it. You'll get there.

Shout outs serve as reassurance from one community member to others, to acknowledge and validate one another's struggles and identities, even if there is no direct exchange between individuals. The shout outs are essentially broadcast to anyone who fits into the identities being mentioned, so they know that others want them to feel recognized and accepted.

The sharing of common interests, experiences, and solidarity not only create a platform for the reclamation of marginalized identities, but also for activism. Activist references contributed somewhat (8%) to the community category content. Social media can create a space for discourse around queer issues for those who may not otherwise have access to relevant politicized information or commentary. Multiple topics in digital interactions can instill interest in collective issues. Collective issues arise as the basis for conversation about connections with family, friends, and society. Such conversations provide the means for gaining insight on the circumstances of others, which can spark interest in political causes, and ultimately allow for a collective, active challenging of societal heteronormativity (Bennet and Segerberg 2011). This dynamic can be illustrated using one of the elements in the data: A photo of a man standing in front of the United States Capitol Building holding a tambourine, a gay pride flag, and a protest sign with big bold letters reading "roses are red, grass is green, never underestimate the power of a faggot with a tambourine." In digital space this photo was turned into a meme, indicating the resonance it held for the queer community. The impact is not confined to digital space. The man in the photo is a New York City activist who was attending the January 21 *Women's March on Washington DC* in 2017 whose campaign proceeds help fund the New York City Gay Men's Chorus in their fight against homophobia and intolerance through song (Bonfire 2017). Digital-space community and content sharing are being combined to expand and challenge the discourse around activism and human rights with material results in the directly inhabited, physical environment.

Most of the analyzed digital content pointed to positive, collective aspects; unexpectedly, however, a small but significant proportion (13%) addressed problems within the queer community. These items either addressed the lack of true community or criticized how queer individuals treat one another in intimate relationships. Consider these two examples: The image of someone simultaneously wearing two different sets of high heels with the caption, "When queer community fails you but femme community builds you back up." The image of actress Nicole Richie (known for her very dysfunctional relationships with other celebrities) smoking a cigarette alongside a bottle of alcohol with this text:

> Watching people you consider to be your friends continue to a associate with people who abused you and knowing there's nothing you can do about it because the queer community is so small

> that every connection we make is important, so you have to stand by and wait for them to get hurt the way you did before they back away.

Small social circles with highly charged common issues can become very intimate collectives with tight-knit relationships among members. When combined with problematic coping strategies, the outcome is not always positive and beneficial.

Marginalized communities do not exist without their divisions. *Respectability politics* refers to efforts made by marginalized persons to pressure others in their group to conform to particular appearance and behavior standards from the mainstream. This can divide the group as they try to determine how members of the marginalized community need to live in order to earn adequate respect from the mainstream. This dynamic gives rise to gatekeepers and, ultimately, the dominance of one social group over others, even ostracizing some members. Gatekeeping expands across race, gender, and other intersections within the queer community. This leads to fractures across the communities in both digital and physical space.

After the legalization of same-sex marriage in 2015, respectability politics was most often practiced by conservative, gay, white men. This was evident in multiple ways, including their attempt to "Drop the T" from the LGBT acronym (Brydum 2015). Their form of respectability politics also embraced racism with "... Gay white men proudly stating their dating 'preferences,' such as 'no blacks' and 'no Asians' " (Williamson 2015), closed-mindedly lumping all people of a race into a single, homogenous "unattractive" category. Two memes in the content data illustrate this concept using famed puppet, Kermit the Frog, speaking to himself:

ME: you are not their teacher, they can Google and educate themselves
ALSO ME: hit them with that 500 word essay on respectability politics.

> Racism and bigotry from within the community make growth and solidarity impossible, and those belonging to multiple intersections are silenced all over again.

Conclusion

Overall, the content analysis supported previous research regarding identity formation, self-presentation, mental health, and digital community among queer individuals. However, three surprising themes emerged. The first, was how overtly members shared that they were deeply aware of their identities and how identity is shaped by others' behavior. The amount of philosophical insight that went into contemplating the self and the expression of that self was unexpected. The second concerns issues with the safety of digital community support. Some elements meant for a specific community were shared among queer blogs, expanding exposure. The number of people who connected with the message being sent was problematic. This suggests that the concept of online community is not always a safe space, and that some individuals may benefit from some time away from the community. The last, and most important, is that the humorous quality of each meme seems to be a coping mechanism *in itself*. Regardless of how morbid or dark the meme was, humor could be detected, which implies that sharing content, anonymous or not, allows space for healing from the hardships of the queer experience.

This research was conducted through the lens of queer theory, which examines the constitutive discourses of homosexuality (and others) in order to place "queer" within its historical context (Jagose 1997:6). Queer theory's main purpose is to frame LGBTQ+ issues as their own issues, rather than against a backdrop of heterosexuality. In general, such work has primarily focused on lesbians or gay men, and how their sexualities are controlled or disciplined in patriarchal or heteronormative ways (Brown 2012:542–543). In the current project, non-binary individuals were mentioned in the general shout outs; however, none of the analyzed content was related to non-binary methods of expression. Asexual and pansexual individuals were also invisible. This research was intended to be inclusive of all identities; however, it seems as though those who were posting the online content often forgot most of the letters in the LGBTQ+ acronym.

References

Allen, Samantha. 2015. "LGBT Leaders: Gay Marriage is Not Enough." *The Daily Beast*, June 26. Retrieved February 17, 2017 (http://www.thedailybeast.com/articles/2015/06/26/same-sex-marriage-is-legal-now-what.html).

Bagozzi, Richard P. and Uptal M. Dholakia. 2002. "Intentional Social Action in Virtual Communities." *Journal of Interactive Marketing* 16(2):2–21.

Bennett, W.L. and Segerberg, A. 2011. "Digital Media and the Personalization of Collective Action: Social Technology and the Organization of Protests Against the Global Economic Crisis." *Information, Communication, & Society* 14(6):770–779.

Berzonsky, Michael D. 2005. "Ego-identity: A Personal Standpoint in a Postmodern World." *Identity: An International Journal of Theory and Research* 5(2):125–136.

Birkett, Michelle, Michael E. Newcomb, and Brian Mustanski. 2015. "Does It Get Better? A Longitudinal Analysis of Psychological Distress and Victimization in Lesbian, Gay, Bisexual, Transgender, and Questioning Youth." *Journal of Adolescent Heath* 56(3):280–285.

Bonfire. 2017. "About This Campaign." Retrieved February 22, 2017 (https://www.bonfire.com/power-of-a-faggot-with-a-tambourine/).

Bostwick, Wendy B., Carol J. Boyd, Tonda L. Hughes, Brady T. West, and Sean Esteban McCabe. 2014. "Discrimination and Mental Health Among Lesbian, Gay, and Bisexual Adults in the United States." *American Journal of Orthopsychiatry* 84(1):35–45.

Brown, Michael. 2012. "Gender and Sexuality: Intersectional Identities." *Progress on Human Geography* 35(4):541–550.

Brydum, Sunnivie. 2015. "LGBT Groups Respond to Petition Asking to 'Drop the T'." *The Advocate*. November 6. Retrieved December 12, 2017 (https://www.advocate.com/transgender/2015/11/06/lgbt-groups-respond-petition-asking-drop-t).

Burnham, Douglas, and George Papandreopolous. nd. "Existentialism." *Internet Encyclopedia of Philosophy*. Retrieved February 4, 2017 (http://www.iep.utm.edu/existent/#H4).

Campbell-Sills, Laura, Sharon L. Cohan, and Murray B. Stein. 2006. "Relationship of Resilience to Personality, Coping, and Psychiatric Symptoms in Young Adults." *Behavior and Research Therapy* 44(4):585–599.

Carver, Charles S., Michael F. Scheier, and Jagdish Kumari Weitraub. 1989. "Assessing Coping Strategies: A Theoretically Based Approach." *Journal of Personality and Social Psychology* 56(2):267–283.

DeAngelis, Tori. 2002. "New Data on Lesbian, Gay, and Bisexual Mental Health." *American Psychological Association* 33(2) (www.apa.org/monitor/feb02/newdata).

Duguay, Stefanie. 2016. " 'He Has a Way Gayer Facebook than I Do:' Investigating Sexual Identity Disclosure and Context Collapse on a Social Networking Site." *New Media Society* 18(6):891–907.

Gable, Philip A. and Eddie Harmon-Jones. 2013. "Does Arousal Per Se Account for the Influence of Appetitive Stimuli on Attentional Scope and the Late Positive Potential?" *Psychophysiology* 50(4):344–350.

Giles, David C. and Julie Newbold. 2011. "Self- and Other-Diagnosis in User-Led Mental Health Online Communities." *Qualitative Health Research* 21(3):419–428.

Goffman, Erving. 1956. *The Presentation of Self in Everyday Life*. New York: Anchor Books.

Grzanka, Patrick R. and Emily S. Mann. 2014. "Queer Youth Suicide and the Psychopolitics of 'It Gets Better'." *Sexualities* 17(4):369–393.

Jagose, Annamarie. 1997. *Queer Theory: An Introduction*. New York: NYU Press.

Luthar, Suniya S. and Dante Cicchetti. 2000. "The Construct of Resilience: Implications for Interventions and Social Policies." *Developmental Psychopathology* 12(4):857–885.

MentalHealth.gov. 2017. "What is Mental Health?" August 29. Retrieved June 28, 2018 (https://www.mentalhealth.gov/basics/what-is-mental-health).

Meyer, Ilan H. 2003. "Prejudice, Social Stress, and Mental Health in Lesbian, Gay, and Bisexual Populations: Conceptual Issues and Research Evidence." *Psychological Bulletin* 129(5):674–697.

Michikyan, Minas, Kaveri Subrahmanyam, and Jessica Dennis. 2015. "A Picture is Worth a Thousand Words: A Mixed Methods Study of Online Self-Presentation in a Multiethnic Sample of Emerging Adults." *Identity: An International Journal of Theory and Research* 15(4):287–308.

Miller, Brandon. 2016. "A Computer-Mediated Escape from the Closet: Exploring Identity, Community, and Disinhibited Discussion on an Internet Coming Out Advice Forum." *Sexuality & Culture* 20(3):602–625.

NCTSN (The National Child Traumatic Stress Network). 2015. "LGBTQ Issues and Child Trauma." Retrieved March 10, 2017 (http://www.nctsn.org/sites/default/files/assets/pdfs/safe_spaces_safe_places_flyer_2015.pdf).

Negru-Subtirica, Oana, Eleonora Ioana Pop, Koen Luyckx, Jessie Dezutter, and Michael F. Steger. 2016. "The Meaningful Identity: A Longitudinal Look at the Interplay Between Identity and Meaning in Life in Adolescence." *Developmental Psychology* 52(11):1926–1936.

Ozer, Emily J., Suzanne R. Best, Tami L. Lipsey, and Daniel S. Weiss. 2003. "Predictors of Posttraumatic Stress Disorder and Symptoms in Adults: A Meta-Analysis." *Psychological Bulletin* 129(1):52–73.

Perone, Angela K. 2014. "The Social Construction of Mental Illness for Lesbian, Gay, Bisexual, and Transgender Persons in the United States." *Qualitative Social Work* 13(6):766–771.

Pew Research Center. 2015. "Section 1: Changing Views of Same-Sex Marriage." Retrieved February 17, 2017 (http://www.people-press.org/2015/06/08/section-1-changing-views-of-same-sex-marriage/).

Reich, D. Bradford and Mary C. Zanaraini. 2011. "Sexual Orientation and Relationship Choice in Borderline Personality Disorder over Ten Years of Prospective Follow-Up." *Journal of Personality Disorders* 22(6):564–572.

Schreiber, Katherine. 2016. "Why Transgender People Experience More Mental Health Issues." *Psychology Today*December 6. Retrieved September 1, 2018 (https://www.psychologytoday.com/us/blog/the-truth-about-exercise-addiction/201612/why-transgender-people-experience-more-mental-health).

Sinclair, Samantha, Richard Carlsson, and Fredrik Bjorklund. 2016. "Getting Along or Ahead: Effects of Gender Identity Threat on Communal and Agentic Self-Presentations." *Scandinavian Journal of Psychology* 57(5):427–432.

Tyler, James M., Peter O. Kearns, and Miranda M. McIntyre. 2016. "Effects of Self-Monitoring on Processing on Self-Presentation Information." *Social Psychology* 47(3):174–178.

Williamson, Kit. 2015. "The Rise of the Gay Bigot." *The Advocate*November 13. Retrieved December 12, 2017. (https://www.dvocate.com/commentary/2015/11/13/rise-gay-bigot).

5

DIGITAL DEPENDENCY INTERRUPTED

Profiles of Withdrawal for Self-Described Internet Addicts

Grzegorz Hryniszak

For Your Consideration

> What a strange practice it is, when you think of it, that a man should sit down to his breakfast table and, instead of conversing with his wife, and children, hold before his face a sort of screen on which is inscribed a world-wide gossip! (Cooley 1909:83)

What does it feel like to forego digital presence for a day? Could you, and would you, voluntarily give up your smartphone, laptop, and tablet for twelve consecutive waking hours? And, why would you even want to do such a thing?

Perhaps you have done it, or perhaps you have forgotten your phone, run out of battery power, or spent prolonged time in a no Wi-Fi zone, but if not, imagine yourself turning off all digital devices, disconnecting from all social media, and yet continuing to function in this physical world. If you are like the participants in the current study—millennial generation college students—chances are you would find such conditions annoying. You might feel anxious and restless, bored, and dissatisfied with your usual everyday activities. You might get angry. There is also a good chance that you might feel isolated and lonely.

The Unbearable Torture of Disconnect

This chapter examines narratives from a project in which 420 millennial participants gave up their access to digital space for twelve consecutive waking hours (on a day and start time of their own choosing) and wrote hourly journal entries reporting on what it felt like to be "offline." This section starts off with a large number of brief quotes to set the tone of the later analytical discussion. Since no specific questions were asked, participants were free to describe their emotional states and behavioral choices in any way they chose. Quite a few narrated their time without digital presence in dramatic terms, clearly demonstrating the great difficulty of accomplishing the task. For many, the experiment turned out to be "the hardest 12 hours [they] have ever had to endure." To some, it was the only half-day period they could remember ever having spent without the ability to instantly connect to social media. In fact, for a significant number of Millennials, the task turned out to be impossible to complete. Close to a third (31%) of all participants were not able to stay away from their devices, and reported "cheating" one or more times. Students wrote about "the torture" of "the eternal twelve hours," which they found "unbearable." In a "dark hour" one wrote:

> I felt like giving up at this moment and just checking my phone. I had nothing to do, nowhere to go, and no one to talk to. Even my family started to notice me pacing back and forth and writing in my journal some bad things about the assignment. I was going crazy.

Another person revealed: "By the tenth hour I was doing a complete countdown, first by every thirty minutes, then by every fifteen, and finally by every five minutes, until I could check and use my phone."

The following direct quotes from their journal entries, provide a sense of what most participants were experiencing:

> My day felt never ending.
>
> By the evening hours it was unbearable.
>
> I was dying without seeing what was going on.
>
> I really felt like I wanted to just yell.
>
> I feel so lost, as if I could not breathe.
>
> Giving up my phone was HORRIBLE. I feel naked, lost, empty and out of the loop. It's a horrible feeling.

One of the journal entries clearly describes a panic attack:

> While being unplugged I felt anxious, nervous, and disconnected from the world. For instance, there was a moment while driving alone that I felt anxious, my heart started racing and I felt short of breath, just because I needed to talk to someone on the phone.

Approaching the end of the torturous assignment, some students wrote about their moods:

> My mood is: anxiously waiting for it to be midnight to speed back to my house and check my phone.
>
> I did not feel OK. My stress level was elevated. I needed to check my emails, but I could not. My husband tried to relax me by distracting me with random conversation. However, my mind was somewhere else.
>
> I have 2 more hours and I am already feeling happy. I feel like it was years that I have not used a cell phone.
>
> I feel happy. I made it the 12 hours without using my phone and I will soon be checking my phone. I cannot wait to check my Instagram and see my text messages. I feel relaxed and relieved.
>
> I feel great. In just a few minutes I will be connected to the social media and I will feel fully alive.

Some final thoughts on their digital abstinence included:

> I went into the idea of being "unplugged" as probably the hardest thing I was going to have to deal with other than a death in my family. It's very sad to say that but it's the reality of how attached I am to a hand-held device.
>
> I started to notice how much my phone controls me.
>
> My phone had taken over my life.

Over 100 similar statements were found in the journals. The picture emerging from these narratives depicts Millennials finding it close to impossible to withdraw from the use of digital technologies, particularly those that are necessary for accessing social media and instant messaging. Many participants reported experiencing strong cravings to use digital devices (65%), particularly their smartphones (60%). The emotional impact of the twelve-hour digital abstinence included feelings of heightened anxiety (63%), annoyance (65%), boredom (53%), dissatisfaction with everyday activities (38%), worry about missing out on social media events (31%), a sense of being disconnected the from the larger world (28%), loneliness and isolation (23%), anger at the assignment (21%), a sense of being incomplete without maintaining their digital presence (18%), and feeling lost without their devices (16%). Millennials also reported feeling depressed (17%) or sleepy (17%), a well-known symptom of depression. Despite explicit prohibition, some could not control themselves and napped (7%).

The data make it clear that within the short time frame of twelve hours, a significant subset of the participants reported cravings and emotional states indicative of compulsive need to use digital technology to connect with others. Many of them used language eerily reminiscent of narratives found among people who suffer from drug, alcohol, and nicotine addictions, as well as those who describe compulsive behaviors, such as gambling.

Students portrayed themselves as "dependent" or "addicted" to smartphones and personal computers:

> I am extremely addicted, hooked and extremely dependent on my electronics like most young people.
>
> During my 12 hours off the grid it was strange to me to find out I was going through withdrawals from not checking my phone for messages or from not picking it up to check the time. I had to constantly keep myself occupied with other things in order for me not to try to reach for my phone.
>
> I'm feeling very anxious and writing this very quickly because as soon as I'm done with it I'm going to run and grab my phone and never let go of it again!
>
> I used to have the tendency to check my Facebook every day. It felt like I had to do it not only once a day but many times through the day. It was like an addiction and my day did not feel complete until I logged on.
>
> By the eleventh hour I feel like the assignment was not only messing with my life but was making me seem like an addict without a drug. I felt a lot more anxious because I knew I was an hour from the finish line.
>
> At this point I've gone 6 hours without the use of my phone. To be completely honest I am starting to despise this assignment. I hate the fact that I had to give up my phone. I feel so agitated. I've never felt this way before. I'm starting to think that I am addicted to my phone. It's the only logical reason that I can think of to explain why I feel this way.
>
> Prior to doing this assignment I would tell my mother that I was not addicted to my cell phone or that I did not rely too heavily on technology like she believed. However, after about six hours into the assignment I realized that she was completely correct. I have become dependent on Google and social media to occupy my time. In order to complete this assignment successfully I had to give my mother all of my digital devices to hide from me so that I would not break and use them. I came to find out that I am most heavily reliant on my cell phone, which gives me access to all my social media sites, email, phone and text messaging capabilities all while on the go.
>
> My dependence on digital things has reached an all-time high and being without them almost drove me into an anxiety attack and even led to me fighting with my wife.

There were participants who, with passage of time, became seemingly incapable of writing longer statements and resorted to one-line declarations, such as: "I feel sick," "I felt sick, better yet, I felt like an addict," or "I realized that not having my phone was the worst thing ever" and "This assignment made me realize I am addicted to my phone." In fact, one-quarter (24%) of all participants described themselves as "addicted" to digital technologies.

Many students reported "feeling naked" and "feeling empty without [the] phone." Some wrote that they "had feelings of emptiness and anxiety," or "kept looking for the phone in my pockets and around and then had to remind myself it was not there and off limits" or "felt odd not having the phone," which "literally felt like [they] were missing something." A few people felt their phones vibrating in their pockets, or heard rings and dings, despite the fact that they had actually left their devices at home. They experienced "phantom phone signals," which existing studies identify as auditory hallucinations (Tanis et al. 2015; Kruger and Djerf 2017). Participants also reported carrying their phones with them throughout the entire day, putting them on tables in front of them while eating meals, and even touching them for comfort, despite keeping the devices turned off.

About two-thirds (65%) of participating Millennials expressed serious, extreme cravings for one or more digital devices, generally their smartphones (60%). What explains such extraordinary attachment to these digital devices? Perhaps it is not as alarming as it sounds; participants could, of course, be using the language of addiction metaphorically, rather than truly expressing their experienced states. Should we be concerned when we hear the following type of declaration? "I am beginning to feel desperate … I definitely feel like a drug addict going through withdrawals!" If reported in relation to compulsive behaviors or substance dependencies, such statements would definitely represent a red flag to any attending physician or psychologist. The next section briefly examines our current understanding of addiction, and how it may relate to the issues associated with withdrawing from the use of digital technologies.

What We Know

Research and theoretical literature on the topic of addiction spans medicine, psychology, and sociology. Medical literature on addiction abounds. It focuses on understanding the physiological mechanisms related to biochemical interactions of addictive substances within the bodies of patients. Nestler describes this biological process as "the ability of repeated exposure to a drug of abuse to induce changes in a vulnerable brain that drive the compulsive seeking and taking of drugs, and loss of control over drug use, that define a state of addiction" (2013:431). *Behavioral addiction*, that is the uncontrolled compulsion to overindulge in activities such as gambling, eating, shopping, exercising, or sex, have also been shown in clinical studies to result in similar pathological changes to patients' brains (Olsen 2011). Psychology, in addition to the pharmacology of substance abuse, considers behavioral and cognitive patterns implicated in developing, sustaining, and cessation of an addiction. It examines initial reasons for taking up an addictive substance or behavior, and how it becomes the source of pleasure that gradually becomes indispensable to an individual. The sociological perspective expands those models by attempting to understand addicts' perceptions of addictive agents within their wider social lives.

In the search for common understandings of what constitutes addiction, Sussman and Sussman explored substantial literature on the subject, compared definitions, and arrived at a set of indicators most frequently included in medical, psychological, and social science research (2011:4025). These are: "engagement in the behavior to achieve appetitive effects, preoccupation with the behavior, temporary satiation, loss of control, and suffering negative consequences." Mosby's Medical Dictionary describes addiction as "A compulsive, uncontrollable dependence on a chemical substance, habit or practice to such a degree that either the means of obtaining or ceasing use may cause severe emotional, mental, or physiological reactions" (2016:37). Beard, who looks at assessing potentially compulsive use of the Internet, believes that a person is addicted when their psychological state (including both mental and emotional states, as well as their scholastic, occupational, and social interactions) is impaired by the overuse of the medium (2005). Thus, the trajectory of addiction begins with individuals seeking out reward stimuli in order to satisfy a craving, continues through temporary appeasement, and leads to repetition of those behaviors despite adverse consequences.

The issue of whether behaviors, such as compulsive gambling or excessive Internet use, should be categorized similarly to classically-defined addictions to pharmacological substances has been debated in a number of fields. Prior to the publication of the fifth edition of the Diagnostic and Statistical Manual of Mental Health (DSM-5) much has been argued both for and against the inclusion of behavioral compulsions, also known as *impulse control disorders*, as diagnosable addictions (for example, Block 2008). Whereas DSM-5 in its final form does contain such conditions as hoarding disorder or gambling disorder, as well as Internet gaming disorder (in the "Conditions for Further Study" section), other technology-related problematic behaviors have not been included.

Another major diagnostic manual, the International Statistical Classification of Diseases and Related Health Problems, Tenth Edition (ICD-10), which is maintained by World Health Organization, also omits technology-related addictions (WHO 1992). It does, however, define impulse control disorders as "characterized by the inability to control behavior despite negative consequences" and the International Advisory Group for the Revision of the ICD-10 (2011) is considering the inclusion of "computer gaming disorder" in ICD-11. Both DSM-5 and ICD-10 explain that there has not been enough clinical research conducted to date to warrant inclusion of other technology-related problematic behaviors. Currently available for discussion, ICD-11 Beta Draft (WHO 2018) proposes to characterize gaming disorder by "a pattern of persistent gaming behavior" with impaired control over "onset, frequency, intensity, duration, termination, context" and "increasing priority given to gaming to the extent that gaming takes precedence over other life interests and daily activities" despite negative consequences. The behavior pattern would need to be "of sufficient severity to result in significant impairment in personal, family, social, educational, occupational or other important areas of functioning" (WHO 2018).

Researchers, mainly in Asia and Europe, have published hundreds of studies (Petry and O'Brien 2013) that argue for inclusion of Internet gaming addiction in DSM-5 and now in ICD-11. Numerous research papers attempted to classify and develop diagnostic criteria for the proposed disorder (e.g. Choo et al. 2010; Fu et al. 2010; Hur 2006). Other researchers dispute the assertion that Asian and European studies provide sufficient evidence, classification, and data (Bean et al. 2017; Petry and O'Brien 2013). Further, Bean and his colleagues (2017) contend that Internet and other technology-related compulsive behaviors are currently seen through the lens of moral panic, which they believe explains the ongoing pressure from some scientific circles about the validity of technology-driven compulsions. *Moral panic theory* explains a dynamic in which populations identify what they perceive to be a threat to the social order, construct narratives about it, then demand that media, politicians, or the scientific community "do something" about the perceived threat (Cohen 1972).

In summary, the only behavioral addictions officially acknowledged in diagnostic manuals are gambling disorder and Internet gaming disorder. Of course, that does not diminish the current study participants' expressed concerns and self-reported realizations about the compulsive or addictive nature of their relationships with their digital devices. The emerging field of digital sociology is very much interested in human-technology interactions; participant journal narratives provide rich data that can be compared to currently sanctioned addiction knowledge.

How It Compares

In 1996, Ivan Goldberg, M.D. published a satirical note on PsyCom.Net (which he founded). In it, he created a mock DSM-IV diagnostic description of Internet Addiction Disorder (IAD). To his great surprise, several of his colleagues in the field of psychiatry honestly admitted to being Internet addicts (Wallis 1997). Ever since, discussions of technology-related addictions consist of serious scientific inquiry, anecdotal evidence, and a fair amount of mockery. After all, can we talk about "television addiction" or "automobile addiction?" Goldberg provided a mirror, in which some scientists and patients saw themselves reflected—a mirror which revealed in roughly hewn shapes who we might potentially become vis-à-vis our technologies if we are not careful enough.

In order to allow for deeper reflection, this section presents the millennial journal data as it can be seen to align with diagnostic criteria for substance and behavioral addictions. Certainly, it makes sense to explore such a comparison between clinical criteria for established addictions to emerging and oddly familiar symptoms of reported, and yet unclassified, compulsive behaviors.

DSM-5 (APA 2013) and ICD-10 (WHO 1992) diagnose *substance dependence* as a condition wherein three or more of the following criteria occur in a period of twelve months:

- Evidence of tolerance (a need for markedly increased amounts of substance in order to achieve effects originally produced by lower doses).
- Withdrawal state when substance use has ceased or has been reduced (evidenced by characteristic withdrawal syndrome for the substance and use of the same substance with the intention of relieving or avoiding withdrawal symptoms).
- Substance often used in larger amounts or over longer periods than intended.
- A strong desire or a sense of compulsion to take the substance.
- Persistent desire or unsuccessful efforts to cut down use.
- Important social, occupational, or recreational activities are given up or reduced because of substance use.
- Substance use is continued despite knowledge of having a persistent or recurrent physical or psychological problem likely caused by substance.

The proposed DSM-5 (APA 2013), and to some extent the beta version of ICD-11 (WHO 2018) diagnostic criteria for gaming disorder, describe the disorder as persistent and/or recurrent engagement in games, often with other players, leading to clinically-significant impairment or distress as indicated by five (or more) of the following in a twelve-month period:

- Preoccupation with Internet games (thinking about previous gaming activity or anticipating playing the next game; Internet gaming becomes the dominant activity in daily life).
- Withdrawal symptoms when Internet gaming is taken away (such as irritability, anxiety, or sadness, but there are no physical signs of pharmacological withdrawal).
- Tolerance (the need to spend increasing amounts of time engaged in Internet games).
- Unsuccessful attempts to control the participation in Internet games.

- Loss of interests in previous hobbies and entertainment as a result of, and with the exception of, Internet games.
- Continued excessive use of Internet games despite knowledge of psychosocial problems.
- Deceiving family members, therapists, or others regarding the amount of Internet gaming.
- Use of Internet games to escape or relieve a negative mood (e.g., feelings of helplessness, guilt, anxiety).
- Has jeopardized or lost a significant relationship, job, or educational or career opportunity because of participation in Internet games.

Of course, addiction and compulsion are individual matters and should be considered case-by-case (with a professional) in any attempt to diagnose such a condition. Mere alignment of a number of proposed DSM-5 and ICD-11 criteria with corresponding elements of student narratives is thought-provoking, but by no means constitutes scientifically-compelling evidence.

The millennial journal project involved temporary withdrawal from all engagement in digital presence. It was meant to illustrate, to participating students, that social conditions constantly evolve and that our social environment includes ever-advancing technologies, which often influence our lives in ways not experienced by earlier cohorts. The idea was also to show that digital technologies replaced older, analog technologies, and that, with a little effort, one can employ those analog technologies to accomplish their usual daily tasks. It was meant to illuminate social context in the light of history—a classic example of what C. Wright Mills termed "sociological imagination" (Mills 1959/1976). Whereas older students who grew up in "analog times" found the experiment not too challenging, and even pleasant, the millennial cohort discovered, often with surprise, that they were not able to function in the prescribed conditions. Those aged 18–34, quickly realized that their lives depended on digital technologies to a large extent. A significant number of millennial participants (41%) volunteered observations on how much they relied, or depended on, digital technologies to accomplish everyday tasks.

The vast majority of participants complained about emotional withdrawal symptoms, such as frustration, irritability, impatience, anger, anxiety, and depression; all aligned with the withdrawal symptoms of behavioral addiction, such as the proposed Internet gaming disorder. During the twelve hours they had to forgo digital presence, a subset of participants persistently thought of their digital devices, missed them, and planned how they would use them, once the assignment was over. This evidence of preoccupation with forgone activities also aligns with diagnostic criteria for behavioral dependency.

Slightly more than one-third (36%) of participants were not able to come up with any alternative activities, or had no interest in analog substitutes. For some, time felt slowed down and boredom set in, as the following excerpts describe:

> Instead of using the technology I pretty much stared at the ceiling in my room.
>
> Time is going by slower than I thought. If anything, this is having a bad impact on me. I cannot seem to forget about my phone or the Internet. I am so used to checking my phone every 30 minutes or less.
>
> 9 am: I just got up, brushed my teeth, and went back to bed. I stayed awake staring at my ceiling until I took another look at my clock and it showed 09:57.

Without their digital devices over one-third (38%) of participants were dissatisfied with everyday activities, which normally would include some digital technology. They talked about losing interest in activities that did not include digital technology components, such as walking the family dog, cleaning their rooms, playing with younger siblings, or sitting around the table and talking to their families.

As mentioned earlier, almost one-third (31%) of Millennials in the study were not successful in controlling their abstention from digital technologies—another diagnostic criterion for behavioral dependency. To a lesser extent, Millennials mentioned developing an ever-increasing need to frequently use their smartphones to check on social media, email, and text messages. One person wrote:

> I noticed that I became dependent on it. Whenever I got bored or had extra time, I was using my phone. I began using social sites that I never used before and only started because I have a smart phone now. I began texting people since I never could before. I constantly check my phone now to check and see if I get any notifications. In such a short period, I have become addicted to my phone.

Participants (29%) confessed to realizing, thanks to the assignment, that digital devices and, in particular their smartphones, constitute a constant distraction, which unnecessarily takes away their valuable time. They noticed that their study time was more productive (18%), and their family interactions improved (27%), while not being able to use their devices. Many students became aware of negative consequences of overusing their phones, computers, and tablets. In fact, the assignment had an unintended outcome in that one-third of the Millennial participants (32%) reported gaining greater awareness of the negative impact digital technologies could have on their lives.

The above comparison of diagnostic criteria for substance dependence and behavioral addictions to symptoms and attitudes described by the study's subjects serves as a spotlight aimed at relatively new digital communication technologies. Since it was smartphones that Millennials missed and craved most, it seems fair to at least consider the possibility that modern cellular phones (which of course are so much more than just phones) can become an agent of addiction. Can one be addicted to a gadget? What is it about those devices that so many find so irresistible?

Bright Dings of Pleasure

It is the domain of sociology to explain the human condition in broader social context. Can groups of people become dependent on a technology or develop some form of addiction that is not substance-linked? If society is an interconnected system of groups having influence on individuals, then channels of communication allow for that influence to be transmitted throughout the system. Our smartphones are no longer relatively-simple devices meant for one-on-one voice communications, but rather pocket computers capable of accessing worldwide communication networks, knowledge databases, and one another. Millennials who were practically born into the world of pocket-size networking devices take the ability to instantly connect to those networks, and to each other, as a given. By design, the twelve-hour digital abstinence project severed that connection and took its young participants into an unfamiliar, awkward, and surprisingly lonely place. That place was simple physical presence devoid of the digital connectivity. It is fair to say that participants gave up more than phones, laptops or tablets; they withdrew from their digital presence and their usual and familiar social networks.

Smartphones particularly serve as portable access terminals to communication networks connecting social groups. To young adults and emerging adults, social media networks are preferred places for community life via digital presence. For any social being, having to leave their community can be stressful. It can trigger emotional responses, such as increased anxiety or sadness. This was verified in an experiment in which short-term smartphone deprivation led to increased anxiety levels, as measured by the State-Trait Anxiety Inventory, an established psychological measurement scale (Cheever et al. 2014). Further, Cheever and her colleagues found the anxiety was more intense among heavy phone users, leading them to speculate that smartphone users may develop psychological dependence. Although Cheever focused mainly on the devices, it is logical to assume, as Twenge contends, that it was not the withdrawal from phones that caused anxiety among experiment participants, but rather the withdrawal from digital spaces that can be accessed through apps and the web (2017b). The concern here is the usurpation of physical relationships; Twenge (2017a) contends that "the number of teens who get together with their friends nearly every day dropped by more than 40 percent from 2000 to 2015; the decline has been especially steep recently."

Millennial journal entries in the current study show that perceived digital technology dependence was more apparent among younger students who had more difficulty completing the assignment, felt more negative emotions, and relapsed more frequently. Journal entries are illustrative:

> In just this first hour my mood is a bit sad and lonely. I feel out of touch with those who I consider closest to me.
>
> Not having your phone is terrible you feel that you are not part of the world. It's sad to say but you feel lost without your phone, I think.
>
> I was not able to communicate effectively which led me to be more secluded and lonely. I also felt lost because I was not up-to-date with my social sites such as Instagram.
>
> I have a need to look at my social networks and feel connected to everyone.
>
> Not having technology in my life for those few hours made me feel like an outcast and as if I was detached from the world. While my friends were texting and using their phones I just felt like I didn't belong … And I did not feel accepted at that moment. Simple things like checking in on Facebook and looking up news on the Internet keep me connected.

Finally, it is important to address the issue of digital presence via social networks replacing physical-world relationships. A number of student journals disclosed that Millennials may actually prefer digital to physical interactions, that they feel more at home on Facebook or Instagram than in directly inhabited, physical environments. One participant wrote:

> Usually, I replace physical interaction with others by being on my phone and removing myself from the outside world …. Technology provides us with the social interaction that we as humans need. On most days, my physical social interaction with others is very limited. I am usually at school or at home and technology, such as my phone and laptop, provides me with the social interaction I need via texting or Facebook messaging my friends.

Social media companies work hard to develop ever more appealing apps. Facebook, Google, and other companies fund research and employ neural scientists, psychologists, and sociologists, aimed at developing techniques and programming that would prolong time spent on their portals and increase frequency of use. For these companies their source of profit is not the end user, but an advertiser; it is important for them to keep their users online for as long, and as frequently, as possible. In such economic arrangements, the user and their attention become goods being sold to advertisers.

A growing number of former industry insiders, programmers, and executives have come out publicly to expose the long-standing practices of social media companies that thrive in the "attention economy." Justin Rosenstein, the programmer who in 2007 developed the "Like" button for Facebook, left a number of social media platforms, limited his use of others, and tries to keep his own children away from establishing an extensive digital presence (Lewis 2017). In a recent interview, he compared Snapchat to heroin and described Facebook likes as "bright dings of pseudo-pleasure" (Lewis 2017). Those bright dings, and similar digital notifications, are largely responsible for the fact that users touch their phones up to 2,617 times per day, mostly prompted by instant notifications from social media websites, email, and text messages (Winnick 2016). As smartphone users respond to those notifications, they become habituated to being simultaneously in digital and physical realms, or *co-present.* Some research suggests that such behavior is conducive to developing dependency and to decreasing cognitive capacity of their brains (Ward et al. 2017).

Social media corporations deny assertions that their services could be addictive, yet the industry, as a whole, continues to explore behavioral psychology, neural science, and sociology in the attempt to make their websites even more appealing to users. Sponsored by LinkedIn, Microsoft, Facebook, and other social media and technology companies, the annual "Habit Summit" was "started in 2013 with the goal of bringing experts from around the world to share best practices and leanings in consumer psychology, design, and behavioral science" (Habit Summit 2018). The April 2018 conference features such events as a presentation by Nir Eyal on "How to Build Habit-Forming Products," a lecture by Gibson Biddle on "Customer Obsession: Satisfying the Unsatisfiable via Consumer Science," and Kees Oomen's talk on "Using Consumer Psychology to Boost Motivation" (Habit Summit 2018). Attendees can learn how to better command users' attention and cultivate habitual use to make their products more successful.

Social media companies also increase participation by encouraging users to actively provide content. In response to a growing body of research findings that viewing Facebook undermines affective well-being (Verduyn et al. 2015), Facebook published its own research (Burke and Kraut 2016) showing that active users who post on the portal report increased well-being, while those who only read, do not. It is worth noting that many researchers who are not funded by social media companies remain unconvinced. Shakya and Christakis maintain that in their study "Facebook was negatively associated with well-being" (2017:203).

So Now What?

This chapter is titled "Digital Dependency Interrupted" because reading and analyzing journals written by Millennials quickly revealed a frequent theme: participants experienced withdrawal symptoms when abstaining from digital technologies that they usually use. Clearly, people now live in two interconnected worlds: the physical space that we directly inhabit, and the digital space of social media and instant messaging. To Millennials, these are not two worlds, but one reality made up of the digital and the physical (Miller 2011). We cannot entertain a possibility of separating the two worlds anymore. They are enmeshed and entangled. It is important to remember, however, that our digital presence depends on the ability to have access to technologies that may or may not be available at all times. Many accounts of digital withdrawal depicted emotional distress, panic, and an inability to cope with physical reality when electronic devices were turned off. Those who relied too heavily on their phone for connectivity found the assignment too difficult to bear.

Digital technologies are here to stay. We as individuals, and together as a society, should proceed with caution. Physical presence should not be sacrificed in favor of digital connectivity. Face-to-face interaction should be cherished—digital connectivity should not require us to supplant interaction in shared physical space. This is not an entirely new conundrum, however. Over 100 years ago, Charles Horton Cooley bemoaned the new practice of morning newspaper delivery as one that enabled men to separate themselves from their families at their breakfast tables with "sort of screen on which is inscribed a world-wide gossip" (1909:83). Today, at the breakfast tables around the world, a similar practice takes place behind computer and telephone screens by many family members. Then, as now, being fully physically present could be sacrificed in favor of transmitted "gossip." As always, we have the option to resist.

References

APA (American Psychiatric Association). 2013. *Diagnostic and Statistical Manual of Mental Disorders: DSM-5*, 5th edition. Arlington, VA: American Psychiatric Association.

Bean, Anthony M., R. Nielsen, Antonius J. van Rooij, Christopher J. Ferguson, and Ronald T. Brown. 2017. "Video Game Addiction: The Push to Pathologize Video Games." *Professional Psychology: Research and Practice* 48(5):378–389.

Beard, K.W. 2005. "Internet Addiction: A Review of Current Assessment Techniques and Potential Assessment Questions." *CyberPsychology & Behavior* 8:7–14.

Block, Jerald J. 2008. "Issues for DSM-V: Internet Addiction." *The American Journal of Psychiatry* 165(3):306–307. Retrieved January 18, 2016 (http://ajp.psychiatryonline.org/doi/full/10.1176/appi.ajp.2007.07101556).

Burke, M. and R.E. Kraut. 2016. "The Relationship Between Facebook Use and Well-Being Depends on Communication Type and Tie Strength." *Journal of Computer-Mediated Communication* 21(4):265–281.

Cheever, Nancy A., Larry D. Rosen, L.M. Carrier, and A. Chavez. 2014. "Out of Sight is Not Out of Mind: The Impact of Restricting Wireless Mobile Device Use on Anxiety Levels among Low, Moderate and High Users." *Computers in Human Behavior* 37:290–297.

Choo, Hyekyung, Douglas A. Gentile, Timothy Sim, Dongdong Li, Angeline Khoo and Albert K. Liau. 2010. "Pathological Video-Gaming among Singaporean Youth." *Annals of the Academy of Medicine, Singapore* 39(11):822–829.

Cohen, S. 1972. *Folk Devils and Moral Panics: The Creation of the Mods and Rockers.* London: Routledge.

Cooley, Charles Horton. 1909. *Social Organization: A Study of the Larger Mind.* New York: Charles Scribner's Sons.

Fu, King-Wa, Wincy S.C. Chan, Paul W.C. Wong and Paul S.F. Yip. 2010. "Internet Addiction: Prevalence, Discriminant Validity and Correlates Among Adolescents in Hong Kong." *British Journal of Psychiatry* 196(6):486–492.

Habit Summit. 2018. Conference Website. Retrieved February 1, 2018 (https://habitsummit.com/).

Hur, M.H. 2006. "Demographic, Habitual, and Socioeconomic Determinants of Internet Addiction Disorder: An Empirical Study of Korean Teenagers." *Cyberpsychology and Behavior* 9(5):514–525.

International Advisory Group for the Revision of the ICD-10 Mental and Behavioral Disorders. 2011. "A Conceptual Framework for the Revision of the ICD-10 Classification of Mental and Behavioral Disorders." *World Psychiatry* 10(2):86–92.

Kruger, Daniel J. and Jaikob M. Djerf. 2017. "Bad Vibrations? Cell Phone Dependency Predicts Phantom Communication Experiences." *Computers in Human Behavior* 70:360–364.

Lewis, Paul. 2017. " 'Our Minds Can Be Hijacked: The Tech Insiders Who Fear a Smartphone Dystopia." *The Guardian*, October 6. Retrieved December 2, 2017 (https://www.theguardian.com/technology/2017/oct/05/smartphone-addiction-silicon-valley-dystopia).

Miller, Vincent. 2011. *Understanding Digital Culture.* Thousand Oaks, CA: Sage.

Mills, C.Wright. 1959/1976. *The Sociological Imagination.* New York: Oxford University Press.

Mosby's Medical Dictionary. 2016. 10th edition. St. Louis, MO: Mosby.

Nestler, Eric J. 2013. "Cellular Basis of Memory of Addiction." *Dialogues in Clinical Neuroscience* 15(4):431–443. Retrieved January 12, 2018 (https://www.ncbi.nlm.nih.gov/pmc/articles/PMC3898681/).

Olsen, Christopher M. 2011. "Natural Rewards, Neuroplasticity, and Non-Drug Addictions." *Neuropharmacology* 61(7):1109–1122. Retrieved January 14, 2018 (https://www.ncbi.nlm.nih.gov/pmc/articles/PMC3139704/).

Petry, Nancy M. and Charles P. O'Brien. 2013. "Internet Gaming Disorder and the DSM-5." *Addiction* 108(7):1186–1187. Retrieved November 14, 2017 (http://onlinelibrary.wiley.com/doi/10.1111/add.12162/full).

Shakya, Holly B. and Nicolas A. Christakis. 2017. "Association of Facebook Use with Compromised Well-Being: A Longitudinal Study." *American Journal of Epidemiology* 185(3):203–211.

Sussman, Steve and Alan N. Sussman. 2011. "Considering the Definition of Addiction." *International Journal of Environmental Research and Public Health* 8(10):4025–4038. Retrieved March 24, 2017 (https://www.ncbi.nlm.nih.gov/pmc/articles/PMC3210595/).

Tanis, Martin, Camiel J. Beukeboom, Tilo Hartman, and Ivar E. Vermeulen. 2015. "Phantom Phone Signals: An Investigation Into The Prevalence and Predictors of Imagined Cell Phone Signals." *Computers in Human Behavior* 51(A):356–362.

Twenge, Jean M. 2017a. "Have Smartphones Destroyed a Generation?" *The Atlantic*, September. Retrieved December 26, 2017 (https://www.theatlantic.com/magazine/archive/2017/09/has-the-smartphone-destroyed-a-generation/534198/).

Twenge, Jean M. 2017b. *iGen: Why Today's Super-Connected Kids Are Growing Up Less Rebellious, More Tolerant, Less Happy–and Completely Unprepared for Adulthood.* New York: Atria Books.

Verduyn, Philippe, David S. Lee, Jiyoung Park, Holly Shablack, Ariana Orvell, Joseph Bayer, Oscar Ybarra, John Jonides, and Ethan Kross. 2015. "Passive Facebook Usage Undermines Affective Well-Being: Experimental and Longitudinal Evidence." *Journal of Experimental Psychology: General* 144(2):480–488.

Wallis, David. 1997. "Just Click No." *The New Yorker*, January 13. Retrieved June 12, 2014 (https://www.newyorker.com/magazine/1997/01/13/just-click-no).

Ward, Adrian F., Kristen Duke, Ayelet Gneezy, and Maarten W. Bos. 2017. "Brain Drain: The Mere Presence of One's Own Smartphone Reduces Available Cognitive Capacity." *Journal of the Association for Consumer Research* 2(2): 140–154. Retrieved December 12, 2017 (http://www.journals.uchicago.edu/doi/full/10.1086/691462).

WHO (World Health Organization). 2018. "ICD-11 Beta Draft. Section 6C71: Gaming Disorder." Retrieved January 30, 2018 (https://icd.who.int/dev11/lm/en#/http%3a%2f%2fid.who.int%2ficd%2fentity%2f1448597234).

WHO (World Health Organization). 1992. *The ICD-10 Classification of Mental and Behavioural Disorders: Clinical Descriptions and Diagnostic Guidelines.* Geneva, Switzerland: WHO.

Winnick, Michael. 2016. "Putting a Finger on Our Phone Obsession." *dscout*, June 16. Retrieved December 22, 2017 (https://blog.dscout.com/mobile-touches).

Part II

MEDIATED RELATIONSHIPS

Preface

These four chapters examine interactions between and among social actors. They focus on the impact of social media and messaging technologies on effective communication, intimacy factors, relationship maintenance, and interpersonal conflict. Being co-present (simultaneously online while interacting with someone offline) recurs as a key issue in three of the four chapters, as does the dynamic of multi presence (when more than one person in the same face-to-face group is engaged in co-presence). In Chapters 6, 7, and 8, a salient issue is how co-presence and multi-presence contribute to ongoing redefinitions of how one "does" relationships with intimates, friends, and families. Another common thread involves the challenges of balancing digital and physical presence—in particular, the challenge of addressing simultaneous expectations of relating to one another offline and in-person, while also being instantly accessible for digital interactions. These three chapters also explore what insights Millennials gained from forgoing digital connectivity for twelve consecutive waking hours (on a day and at a start time of their choosing). Of particular interest are Millennials' post "unplugged" reflections about balancing their co-present and multi-present lives when attempting to maintain intimacy, scaffold friendships, and interact with family. Chapter 9 specifically looks at relationship conflicts that have a digital component. This chapter draws from an analytic review of well-established and popular advice columns, rather than the millennial journals. It identifies the digitally-relevant relationship conflict issues given the greatest visibility, provides concrete examples of how these conflicts play out in everyday life, and discusses these issues in order to enable broader examination of their overall importance.

Whether explicit or implicit, all four chapters have strong foundations in theory. The Part II chapters primarily draw on social psychology and the micro-social institutions of family and peer groups. In this regard, a broad array of subtopics are represented, including but not limited to: intimacy theory, family dynamics, norm violations, re-socialization and digital netiquette. Concepts of normative pressure and social expectations are particularly well-reflected. Other theoretical and conceptual foci are also present, however. These include earlier digital sociology concepts.

Chapter 6 draws from role theory and socialization in addressing the social expectations that attach to the role of "intimate." It looks at how norms for digital relationship maintenance are reinforced and internalized through routine. Social expectations for an intimate are examined (and explained) in view of intimacy theory's components of breadth, depth and self-disclosure, and openness and meaningful sharing. Finally, intimacy-building and the communicative expectations between intimates are compared to the communications dynamics that digital sociologist, Ben Ze'ev, identified as particular to purely online relationships—distance and immediacy, lean and rich communication—all of which are defined and explained within the chapter.

Chapter 7 incorporates the family studies/ethnic studies concept of *familismo*, as defined within the chapter. The discussion addresses the three constitutional elements of familismo—family structure, attitudes about family, and familial behaviors, to determine if this is still a strong Latinx family characteristic. It also examines Latinx family life as it is lived with at least one family member who is a millennial college student. Possible shifts in family dynamics and in the ways in which Latinx Millennials think about, and enact, digital co-presence and multi-presence while "doing" familismo are central. Latinx is the term of preference here because it avoids the gender binary in the use of Latino/a or Latin@.

Chapter 8 does not explicitly identify theoretical premises, but draws from micro-social and broader societal concepts. At the micro-social level, role primacy is a focus. In the absence of other factors, role primacy determines which of the social roles one inhabits "wins" when one has a role conflict (is unable to simultaneously fulfill the demands of both roles) Specifically, Chapter 8 looks at choices made around the conflicting expectations of the digital companion role, and the physically-present companion role. The broader societal concept of cultural hegemony underlies the discussion in Chapter 8. Specifically, elements of shared cultural experience that are so pervasively integrated into normative cultural expectations, behavior, and attitudes, that most members of the culture no longer reflect on these cultural elements as *optional*, but have come to see them as just *the way things are*. Questions that Chapter 8 considers are: When social expectations for digital connectivity compete with the interaction expectations of those physically present, which dominates? When digital presence has role primacy, how are friend, family and other relationship dynamics affected? Is digital primacy part of the millennial cultural hegemony? Given the hegemonic nature of ongoing digital presence, how is the "taken-for-granted" aspect affected for participants who undertook the anti-hegemonic condition of being digitally abstinent?

Chapter 9 implicitly, rather than explicitly, focuses on normative expectations, norm violations, and re-socialization. This chapter identifies how these dynamics are manifested in advice columns, as multiple cohorts learn to renegotiate self-management issues, and relationship expectations, in view of digital space behaviors. Popular cultural notions of *media reflexivity* underpin the chapter. In other words, advice columns are reflexive because they increase the visibility of conflict around social expectations in digital life, thereby enabling discussion and possible re-consideration of such expectations. On the other hand, advice column letters also contribute to the role media plays in potentially reinforcing newer or changing social norms—as agents of *re-socialization*—providing greater visibility of the expectations about, and the rationale for, supporting newer understandings of appropriate behaviors in digital space.

6

HOW DO I LOVE THEE? LET ME TEXT THE WAYS

Interfacing Intimacy

Cristina Bodinger-deUriarte

Texting Background

The first text messages could be sent as early as 1995, but only within the same network. Most people did not use this technology for the first four years it existed; those who did, typically sent less than one text message every two months. This started to change in 1999 when mobile phones were first capable of sending messages between networks, a process that became simpler in 2000. The average user in 2000 still only sent thirty-five text messages per month. The oldest cohort of Millennials had already turned eighteen at this point; the second generation of Millennials, those who were aged thirteen to seventeen in 2009, came into teen-hood having been long aware of texting. They texted at a rate of fifty messages a day, to average 1,500 per month (Lenhart 2012). They were the first cohort to text more than call. In 2010, the *Computer Industry Almanac* reported that nearly three-quarters (72%) of US citizens over three years of age used the Internet (2010). By 2011, teenager use of texting averaged between about 3,000 messages sent and received by males to almost 4,000 messages sent and received by females each month, an increase dubbed the "mobile data tsunami" (Nielsen Newswire 2011). Meanwhile, the first millennial cohort had upped their average texting rate from 35 to nearly 1,000 messages per month.

Data

Data for this project were collected from 420 millennial college students who gave up digital presence of all kinds, with everyone, for twelve consecutive waking hours at a start time of their own choosing, on any single day within an eight-week window. They generally alerted family, friends, and romantic partners of their offline status. Participants kept journals containing hourly reflections on what they gave up, and how they felt. Findings presented here are based on the resulting 5,040 journal entries. Most participants also included a summative reflection as a capstone to their journal entries. This material was treated as anecdotal evidence. Both journal entries and reflections were completely spontaneous.

Texting Norms Among Intimates

This population navigates the balance between physical and digital presence in a way that influences daily rhythms and routines, the loss of which, even for a short time, is deeply felt. In relation to their romantic partners, three types of connectivity were keenly missed:

- Engaging in simple digital presence
- Engaging in multi-presence
- Engaging in multi-present connectivity.

Simple Digital Presence

This type of digital presence is characterized by sending one another digital messages or material *while temporarily apart*—even if shared physical presence just occurred, is about to occur, or is regular and frequent. Participants found it difficult to forego reciprocated digital contact during the portions of the day that they were not physically together. Comments illustrate the intense need to communicate with an intimate on more of an ongoing basis:

> I feel like time has stood still while I wait [for the 12 hours to end] … that my heart is missing.
>
> I would not be able to communicate with my boyfriend. This was a daunting task for me since we are attached at the hip. It's a little embarrassing but sometimes we will be on the phone and not send anything, or if we see each other I will text him right after and ask what he is doing. I was overwhelmed by not having my phone at this point.
>
> Many of my thoughts and ideas were kept inside of me to share, like lava waiting to explode from a volcano. [After one hour.]

These were typical responses. Time away from contact was not the only factor. Digital presence is well-integrated into the daily rhythms of millennial couples' relationships. These comments indicate that digital presence is used to extend and enhance the sense of personal contact within relationships, and is experienced as integral, not merely additional.

> We usually let each other know through text messages and phone calls what we are doing and plans throughout the day.
>
> Shortly after getting back from break, around 10:40 am, my boyfriend texts me every day. I'll sit and do work at my desk and sneak in a few texts for about 30 minutes until he starts work. We do this every single day.
>
> I would have been texting with my boyfriend since this is the time he wakes up. But instead, I called him from my home landline. Which was very weird because we usually always text each other. We would usually only call each other if it was something very important or if it has been a long time since I answered his text messages.
>
> Once I got dropped off at home [by my boyfriend], I called my boyfriend with my house phone to let him know I was already upstairs. This is a routine we have always followed, except this time I called him rather than sending him a text.

It has been suggested that such frequent use will fade with younger cohorts who no longer experience messaging technologies as "new," however, studies show that "most teens in romantic relationships assume that they and their partner will check in with each other with great regularity throughout the day" (Lenhart, Smith, and Anderson 2015).

Multi-Presence

Multi-presence is when two or more people are co-present—being physically present with one another, while being engaged in *separate* digital presence elsewhere. Many described multi-presence as typical of their "wait-time" when on a date. They would be together, while they each accessed digital material (checking messages, social media, etc.), turning attention more to a partner once the activity began in earnest (food arrived, they entered the theater, etc.). Participants found their waiting routines disrupted when they could not engage in multi-presence. Intimates are described as multi-present when *both* individuals are being co-present at the same time, *without sharing their digital activity with one another.*

> I noticed that he would be giggling on his phone, which made me wonder what was so funny. At this point I did begin to miss my phone because typically we each watch funny videos.

Multi-Present Connectivity

In this case, people are physically present with one another, while also *sharing their digital activity*, augmenting their interaction. They may engage in texting one another and mutual friends, or real-time tweeting while watching a program together. Another example is to share videos, messages, and photos from social media sites, punctuated by commenting on them to one another, and so forth.

> After dinner someone usually cleans up and we … turn on the TV … We usually don't talk because we're just so busy on her phone laughing at what people are posting.

When the ability to simultaneously share both physical and digital space was not possible, the shared experience often felt unsatisfying and incomplete:

> We watched the movie but could not tag … on Facebook. It made me feel detached from what was going on. My friends were tagging us on Facebook but I could not.

Is More Better? More of What?

If the new normal is sending 100–140 texts per day, how many of those should intimates expect from one another? The topic arises in advice columns, blogs, the websites of "dating experts," and other online forums. Psychologists also address the relationship issues that arise around texting expectations within couples. According to psychologist Nikki Martinez, "three to five times is ideal for a couple who see each other morning and night, while if you live apart, you may want to text more often" (Blatchford 2017). This "ideal" seems to be at odds with overall texting frequency.

Reddit is a forum where this topic is frequently revisited (2015, 2016). In 2016, the ASKMEN survey question was, "Guys (21+) with girlfriends: how often do you exchange texts?" The issue was the number of texts that would be enough not to seem uncaring, but not so many as to seem clingy. The most commonly agreed-upon sentiment was that: "An exchange of texts is a great thing. A one-sided barrage is not." Aside from this, responses varied but most expressed expecting/preferring multiple texts a day:

> I realized I like texting … I honestly felt neglected, and it started getting into my head because now I felt that I was the clingy one by wanting to hear from her. I especially didn't want to date someone that made me feel like I was inconveniencing them or bothering them by sending a text. It was part of the reason we broke up (Reddit 2016).
>
> We texted I'd wager 6–20ish times a day, and we were in the same dorm building, saw each other once a day if not more often,

> and slept over a few times a week. I got a "good morning" text … a few conversations during the day (my professor is an asshole, I feel kinda sick, great sex last night, go to the CJL the food is good today, etc.), as well as checking in for locations/meet-up potential in the evenings, sometimes in the afternoon, and usually for dinner (Reddit 2016).
>
> … There *is* such a thing as texting too little, and I like to hear from my SO on occasion, unprompted. I still think texting constantly, or routinely (like good morning texts, ugh) is … stupid, but I definitely want to hear from her. Especially about random shit. I made an online friend and we talk about random shit happening during our day, usually every day, and I love it. I want that in a partner (Reddit 2016).

Such informal online survey responses are validated by formal academic surveys that found some of the most popular text content to include: funny stuff; what you are doing, seeing, or eating; where you are; and stuff going on with other friends (Lenhart, Smith, and Anderson 2015). Studies have shown that sharing the "random" text content is seen as a form of "relationship maintenance" and makes people feel closer to one another, like they are extending a sense of being together when physically apart. This was reflected in our journal entries and in Reddit survey responses:

> I took notes in a notebook [during class], and did not check my cell phone. This was pretty difficult for me, because I usually am texting my boyfriend throughout the day.
>
> My girlfriend and I text all the time. The best texts are just funny random thoughts with the sole purpose of making each other laugh. We love each others weird brains and random thoughts. We can have an hour-long conversation about absolute nonsense, but that's why we love each other so much (Reddit 2016).

Melanie Schilling, a successful, entrepreneurial "relationship strategist" says that "Texting has become one of the (many) forms of legitimate communication for couples, and each couple defines their own 'normal' when it comes to communication" (Blatchford 2017). Studies suggest, however, that texting within close relationships has normative expectations that do not reflect the "to each their own" approach Schilling mentions (Coyne et al. 2011; Hall and Baym 2011). Further, whether or nor anyone agrees with the definition of "ideal" as three to five texts a day (Blatchford 2017), this number is far removed from actual practice. Though current studies disagree on the average number of texts between intimates, Hall and Baym found thirty-eight sent/received daily messages to be typical among close friends

(2011:323). They attribute this, in part, to the dynamic identified by Katz (2008) that mobile phone technology enabling "continuous communication" has influenced our choices around frequency—including the perception that quantity matters more than content. Similarly, Hall and Baym assert "that the mere use of mobile phone technology in friendships creates obligations to be available and responsive" (2011:325). These findings concerned close friends rather than intimates; however the frequency expectation among intimates may well be comparable, considering that intimates are held to stricter response times than friends and bosses, and that being at work is not seen as an excuse for an intimate to delay responding (HTC Team 2013).

Does the constant contact offered through digital presence, the level of interaction during multi-presence, and the type of mutual sharing during multi-present connectivity, actually strengthen intimacy, as millennial couples have suggested? One way to test this is to see if anything changes when digital abstinence is introduced into the relationship.

Digital Abstinence

Millennial adults and emerging adults experience digital presence, multi-presence, and multi-present connectivity as relationship norms for couples. Consequently, participants in the current study reported anxiety, and even conflict, resulting from foregoing these aspects of their romantic relationship during this brief period, even after having alerted their partners in advance. In fact, almost one-third (31%) of the participants were unable to remain non-digital for the full twelve hours. Most of the non-compliance (78%) involved checking text messages.

It might be assumed that the absence of digital access for one partner would lead the other partner to forego the usual wait-time routine in order to interact. Overall, however, this is not what happened. Those no longer able to share wait-time multi-presence, or the "togetherness" typified by multi-present connectivity, were generally ignored by their intimates, who continued to enact their familiar, digitally-enhanced routines (63%).

> I wanted to motivate my fiancé to unplug with me at least for this hour. … This had a negative impact on my fiancé. … He was hesitant to put down the iPad.
>
> We sat to wait, but while we sat, I found myself bored because I did not have anything to take my mind from waiting … Fortunately, we only waited for five minutes … typically we are on our phones [while] waiting to be served.

Only rarely did others voluntarily forego their own digital activities during their shared physical presence with the participant (11%). Of the remaining 306 participants, more than a third (37%), often somewhat aggressively, insisted on having others share their own digital deprivation during the time they were together.

> Not being able to use my iTouch impacted my boyfriend because he wanted to use his iPhone and I told him he couldn't because I couldn't use the Internet, so he had to sacrifice his iPhone also.
>
> While at the mall, I asked my girlfriend to participate in this exercise with me, which included mainly refraining from using her mobile phone. Being at the mall, I can tell she was getting frustrated not connecting with her friends.

Does Digital Absence Make the Heart Grow Fonder?

Does simple digital presence affect intimacy-building between romantic partners who frequently share physical presence? In other words, does it increase the *key components of intimacy*: breadth, openness, and depth (Cox 2009)? Or, does some degree of digital abstinence increase relationship intimacy?

Relationship Breadth

"Breadth" is the aspect of intimacy comprised of *time spent together*, and the *range of activities two people share*. Popular conceptions of Millennials include the stereotype that they replace being physically present with being digitally present, eroding true intimacy through prolonged absences and few shared activities. Data do not support this perception. Journal entries show that removing digital access did not increase physical contact; of the 2,540 journal entries that describe alternative activities, very few (1%) sought additional encounters with intimates during digital abstinence. This indicates that Millennials are *not* significantly replacing physical presence with digital presence. It seems couples commit what time they can to sharing physical space, then augment their connection through simple digital presence. This is supported by another study finding that "the more the participants engaged in face-to-face interaction … the higher their motives were and the more frequent cell-phone use was;" in other words the more they shared physical presence, the more it triggered digital presence—loneliness did not trigger higher technology use (Jin and Park 2010:611). Journal entries supported this when talking about digital absence as deprivation:

> My boyfriend arrived home so we decided to get ready and go out for a little date. Not speaking to him all day made me miss him a little more since I did not have contact with him.

Augmenting physical presence through texting, along with other messaging and social media technologies, comes to be seen as part of "being together" and helps explain the 24/7 five-minute response-time expectations for intimates.

On first glance, foregoing digital activity while physically present seems to increase the range of shared activities in two ways: (1) engaging in *novel activities*; and (2) engaging in translated activities that parallel familiar digital activities. Activities are deemed novel when they are unusual and unlikely to be repeated on a regular basis. They arise spontaneously, often in response to unusual circumstances. For example:

> Since we didn't have our phones we constantly felt we needed to do something in order to keep ourselves occupied. So we sat for thirty minutes after lunch and talked and made a fire to eat smores. We had fun actually … making a fire in our fireplace. We sat around the fire and enjoyed each other's company.

Translated activities are simply the non-digital version of a similar pastime. For example:

> Since I could not go on my Ozone YouTube … we decided to sing. This created so much laughter, as we listened to each other make up songs. We began to really enjoy ourselves.
>
> I've never played a card game with him before so it was a new experience. We usually play video games together like Donkey Kong on my retro game emulator or NBA 2K13 on his PS3 … .

There is no indication that these shifts to non-digital activities improve intimacy by contributing to relationship breadth in a meaningful way. Journal entries acknowledge the fun of novel activities, but neither the individual entries nor the concluding written reflections depict these activities as increasing a sense of connection. Instead, creative use of time together seems simply to reflect individual creativity. It is not a common result of being offline.

Data support this interpretation. Of the 2,540 journal entries that describe alternative activities, just over 2 percent (44) mention using creative alternatives when access to digital space is lost. On the other hand, 1,101 (43%) describe alternative activities that are only simple translations of the digital into the non-digital equivalent: gaming, reading, listening to music, shopping, etc. These activities do not represent any significant change in the *range* of shared activities—only in the *way* those activities are shared. Intimates are clearly able to form strong bonds without adding to the activities they share—even if most of the activities they share are online.

Relationship Depth and Self-Disclosure

Depth describes the mutual self-disclosure that gives partners a *true, central, and meaningful understanding of one another.* Digital presence both contributes to this and detracts from it. Online communication has long been shown to facilitate the sharing of deeply personal, emotionally charged, or embarrassing things that intimates feel are important to disclose to one another (Yum and Hara 2005; Ben-Ze'ev 2004). Beneito-Montagut agrees that in his study "some participants report that it is easier to have intimate conversations online than face-to-face ... depending on the topic of conversation, they prefer using IM [instant messaging] over face-to-face options" (2015:546). The following quotes illustrate this:

> When I'm in person with him, like, it's harder for me to tell him what I'm feeling. So, like, I'll think about it when we're together, and then, like, afterwards I'll probably text him, like, what I was feeling and tell him my problems (Lenhart, Smith and Anderson 2015:43).
>
> I think texting kind of makes you feel closer ... my boyfriend, he doesn't like to express himself like that. But when we text, it seems like it's so much easier for him to talk to me. So I think he says more stuff, like how he feels, through text. So it kind of makes [the relationship] stronger (Op. Cit.).

These communications also include serious issues that affect how these young couples understand one another more deeply. For example, "half of teen daters have discussed difficulties with family online with their significant other, while 31% have discussed personal health issues in these venues" and many (70%) "have felt closer to a significant other because of text/online conversation" (Ibid 35, 39).

There is more to the story, however; the same study also shows that almost half (43%) "have encountered a situation in which they felt that their partner was distracted by their phone when the two of them were together in person" (Ibid 40). Journal entries also support this latter observation in participant comments that indicate digital abstinence gains in perceived attentiveness, at times leading to increased understanding of one another:

> My girlfriend was loving all the extra attention I was giving her. She even said, "You should do this project more often." This coming from a girl who is literally on her phone 24/7.
>
> [We] actually talked more than usual and it was actually very nice ... It was very enjoyable just being able to sit with my boyfriend talking about life.

> I enjoyed more conversation with my girlfriend instead of … having my I-Pod … or texting … being able to talk about how she met her friend who had the baby shower and how they helped each other through medical school gave me a better understanding of how difficult medical school was and made me so much more proud of how hard she worked … We were able to have a more personal conversation that helped us understand one another.

Relationship Openness and Meaningful Sharing

Openness describes an *ongoing willingness to share intellectually, physically, and emotionally*. It includes *trusting and being trustworthy*, as well as expressing affection.

This is the area most often described as providing gains during times when physical togetherness did not include digital presence. For example:

> We are so into technology that we forgot how nice it can be without it. We spend so much time apart with school and work, so I was grateful for this opportunity to just catch up and enjoy each others company. We really talked together for about forty-five minutes and then made our way down back to the car.
>
> I'm realizing that I focus more on my girlfriend without technology. What a difference not checking your phone for 4 hours makes! Talking, socializing more, and, in ways, being more intimate with one another.
>
> Instead I had coffee and breakfast with my husband. We had good conversation. It was a very nice change. My mood was very good.
>
> [We] found that without our electronics we talked way more than we had been lately. We caught up about school and work and even made plans for the following week to hang out all day without our electronics.

This type of response was frequent; however, it was limited to describing gains when physically together. Ben-Ze'ev's observed that intimates can become so used to one another's physical presence, they stop practicing true attentiveness when together and potentially undermine intimacy-building. However, Ben-Ze'ev went so far as to describe fully-online relationships as potentially more successful and quicker in intimacy-building, than fully offline relationships (2004).

Intimacy—Digital, Physical, or Both?

Fourteen years ago Ben-Ze'ev identified specific ways that online dating and online-only romantic relationships, differed from relationships in which people interacted face-to-face. He claimed that online relationships had "opposing features" that would not make sense in offline relationships (2004). Opposing features included exchanges that were both distant and immediate and could involve both lean and rich communication. At the time of his research, the few people who sent text messages, only sent one message per day. Millennials now typically send 100 to 140 per day; many of them going to intimates they spend physical time with. In view of this shift, it is worth asking: whether concepts of "distance and immediacy" and "lean and rich" communication opposites still distinguish fully online (digital space only) relationships from other relationships.

Communication Distance and Immediacy

Ben-Ze'ev explains that romantic partners in online relationships occupy the same space at the same time by being online together; this builds emotional and communication "immediacy" because partners focus on one another. With the advent of texting, however, the same communication immediacy can be interjected at any point in the day of the so-called "offline relationship." Journal entries show texting as an extended sense of togetherness, one in which the importance of immediacy is underscored by rapid response-time expectations. This supports findings from Beneito-Montagut's interview-based study that immediacy "is one of the main motivational factors driving online communication" (2015:546). He quotes a respondent: "I will never say, 'let's wait until tonight (to communicate something) as I could see you face-to-face then'." (ibid).

Ben-Ze'ev also depicts online couples as uniquely able to use physical distance to moderate a reply by delaying a response, whereas offline couples are expected to respond instantly when face-to-face.

> People can express their authentic, spontaneous emotional reactions, as is done in offline relationships. However in online relationships, people also have the choice of postponing their reaction, in order to allow time to moderate their response (2004:28).

Texting now affords some opportunity to delay responses, but only for a short time. Response-time expectations have changed significantly since 2004. It is unlikely that online relationships have escaped the escalating expectations around texting response times. One worldwide study using focus-groups and interviews found that, depending on the type of communication, an immediate response could be expected. Although in some cases, ten minutes to an hour might be acceptable, "texting has habituated us to a much quicker response … when we don't get a quick response, our mind freaks out" (Ansari and Klinenberg 2016).

The High Touch Communications Team (HTC Team) conducted a US-based survey on response-time expectations, dividing categories into: (1) email, vmail, and texts; (2) whether it is during work hours or off hours; and (3) who the message is from—boss, colleague, family member, friend, or romantic partner. The response-time expectations for text responses were significantly shorter than for any other medium—by as much as 24 hours. The text response expectations for romantic partners did not change by work status. They were also the most immediate, at five minutes. The HTC Team depicted this as an expectation that romantic partners are "On-Call 24–7" (HTC Team 2013). There is no indication that communication distance and immediacy—and their potential effects on intimacy—differ by the degree to which intimates do or do not share physical presence.

Lean and Rich Communication

When Ben-Ze'ev studied online relationships, he suggested that communication between partners in online relationships was, of necessity, rich—that is more developed and more profoundly communicative—and that "in this sense, less can mean more" (2004:30). His position opposed common perceptions that pictured rich communication as only meaningful through face-to-face exchanges, and in-person "heart-to-heart" talks. Technology has significantly changed, however. Real-time chat functions that include live video are increasingly available (such as Skype, zoom, and other messaging software). This technology undermines the conditions Ben-Ze'ev thought enhanced the ability to "read between the lines" of posted communications.

Rich communication might also include moving highly-emotional conversation into digital space to limit interruptions that derail communication, make sure no points are overlooked in the heat of an exchange, and provide an accurate account. Face-to-face may also be less effective at times because some content may be missed while one is formulating an instant response to what is being said, and total recall is not typical.

As discussed earlier, however, rich communication takes place both in physical and digitally-mediated interaction. The average millennial couple uses technology to combine lean and rich communication with digital exchanges that often involve depth, self-disclosure, openness, and meaningful exchange.

Texting, has been considered a *lean communication* style because it *lacked expressive elements* associated with rich communication involved in face-to-face conversations. Lean communication generally lacks tone of voice, body language, facial expressions, physical contact, and other cues that indicate attitude, emotional state, and visible sincerity. Ben-Ze'ev flipped this idea around. He claimed that the online potential for misunderstanding irony, sarcasm, and "kidding-around," was compensated for by the concentrated focus on one another that messaging promotes. He diminished the importance of physical cues, claiming that those who shared physical presence became so used to one another that they were frequently "blind" to the social cues available. Intimates do not so easily dismiss the importance of greater expressiveness in digitally-mediated communication, however. This is apparent in the variety of apps now marketed to address the absence of communication cues when not face-to-face. Such apps provide for instantly available personal photos, vocal messages, self-filmed video clips, doodles, emojis, avatar-based images (as with bitmojis), emoticons, stickers, animation, and so on.

Affection? We Have an App for That

Certainly marketers have latched onto the concept of digitally-mediated intimacy-building as a lucrative opportunity. This can be seen in numerous apps targeting couples. For example:

> *Ice Break* entails a list of questions to ask one another, and is focused on creating more meaningful talks and on understanding your partner better.
>
> *Couple* provides constant connection on a shared timeline. It allows users to express moods and feelings through the exchange of texts, video, audio, doodles, cute stickers, and even private photos (which can be set to disappear at a given time). Intimate partners can set notifications on a shared calendar and exchange to-do lists.
>
> *Feel Me* shows a small red dot at the place where one partner's fingers are currently touching the screen while composing a text. When the two partners press the screen in the same place, a pleasant vibration occurs.
>
> *You & Me* knows when both of you are together in real life and adjusts the interface. It has special photo formats for couples, including "Secret" where private pictures appear "steamed up." The recipient wipes the screen to view.
>
> *Fix a Fight* suggests proven tactics to "make-up" with one another, and gives advice on how to avoid that fight the next time. It also advises on how to better express your feelings, and initiate discussions about things that bother you.
>
> *LokLok* lets intimate partners exchange funny doodles, sketches, and hand-written notes. These instantly become visible on the partner's screen without unlocking it.
>
> *Hey Tell* features a button to record a message and have it delivered to your partner. It's like a private voicemail account, where you can listen and reply to messages when it's convenient for you.

These are only a few examples of apps being marketed as intimacy-builders for couples. Millennial couples are, indeed, using them. For example, the app *Between* is a platform for couples to exchange and store photos, voice messages, cute stickers, and animated emoticons. In just one year it reached one million downloads. Current users average five hours a month on the app (Prokopets nd). The success of affection-expressing technologies makes sense in view of the fact that Millennials and post millennial teens identified "expressing affection" as the most common reason for digital contact (Coyne et al. 2011; Jin and Park 2010; Schade et al. 2013).

Conclusion

Texting has become an indispensable part of modern relationship maintenance and a powerful means of intimacy-building. Intimates feel greater closeness because they feel included in one another's unfolding day. Internalized relationship norms now include significant expectations for physical presence, simple digital presence, multi-presence, and multi-present connectivity. This is deeply ingrained, as indicated by participant reflections in the current study. Despite insights gained, only a few participants said they would forego their devices again for any period of time (8%). However, about a third of the participants said they now realized how important it was to be *more aware of a balance* between digital presence and physical presence *when together* (32%). Slightly more said that they now realized that *fully focusing on being together* during a significant interval allowed for interactions that were qualitatively different (39%). In short, the importance of moderating multi-presence, and multi-present connectivity were recognized. However, Millennials remained committed to pursuing a sense of ongoing connectivity through combining digital presence and physical presence. This is true even for partners who regularly date or cohabitate.

The current study supports earlier findings that digital exchanges need not always be meaningful to enhance intimacy. In fact, simple "updates, narratives, and short messages sent through mobile media increase interaction and inclusion felt between individuals who are not [sharing physical presence]" (Hall and Baym 2011:320). In this way, lean communication contributes to relationship maintenance. Interestingly, in 2004 Ben-Ze'ev identified this dynamic as a strength for online relationships. It now holds true regardless of how couples integrate digital *and* physical presence into their daily rhythms and routines. Recent research, including the current study, also corroborate Ben-Ze'ev's claim that leaner communication does not necessarily lead to shallower relationships. It has different effects depending on how it is used. "What is important in this regard is the nature of connectivity, not the content" (Ben-Ze'ev 2004:33).

One issue for intimacy-building becomes the effectiveness couples develop in communicating across mediums, and the choices they make about how to communicate. Schilling recognizes this, warning that healthy relationships partly depend on *how* texting is used, and that this may well vary among couples. "For one couple, a text might be viewed as a poor excuse for a conversation, even as an avoidance of an issue. Whereas for another couple, it may be part of the normal flow of communication in their day" (Blatchford 2017). Recent studies find that intimates have a well-developed and wide-ranging use of texting that includes self-disclosure and argument resolution that would otherwise have been seen as impossible to accomplish face-to-face.

A second issue involves the delicate balance between feeling included and feeling obligated. Research has shown that relationship satisfaction is improved when couples have agreed on basic rules about texting and phone use (Miller-Ott, Kelly, and Duran 2012). This is also the best strategy for avoiding accelerating texting expectations. Hall and Baym found that, as digital communication increased, so did expectations about future contact. At some point, the initial satisfaction about inclusiveness shifts to dissatisfaction with the sense of obligation around increasing digital communication (2011:324). These choices will inevitably embrace both immediacy and distance by using digital presence and small response-time delays—as well as embracing both lean and rich communication in digital and physical space.

Overall, findings strongly suggest that digital presence of all kinds can act as an enhancement in intimacy-building, but only when couples also make adequate inclusion of other forms of self-disclosure and connectivity through face-to-face conversation and shared activities that involve more attentiveness to the partner.

References

Ansari, Aziz and Eric Klinenberg. 2016. "She'll Text Me, She'll Text Me Not: The Science of Waiting in Modern Courtship." *Nautilus*, February 11. Retrieved April 2, 2018 (http://nautil.us/issue/33/attraction/shell-text-me-shell-text-me-not).

Ben-Ze'ev, Aaron. 2004. *Love Online: Emotions on the Internet.* Cambridge, MA: Cambridge University Press.

Beneito-Montagut, Roser. 2015. "Encounters on the Social Web: Everyday Life and Emotions Online." *Sociological Perspectives* 58(4):537–553.

Blatchford, Emily. 2017. "How Often Should You Text Your Partner?" *HuffPost*, August 9. Retrieved February 17, 2018 (http://www.huffingtonpost.com.au/2017/09/07/how-often-should-you-text-your-partner_a_23198283/).

Computer Industry Almanac. 2010. *Computer Industry Almanac.* Retrieved April 2, 2018 (http://www.c-i-a.com).

Cox, Frank D. 2009. *Human Intimacy: Marriage, the Family and Its Meaning, 10th edition.* Belmont, CA: Wadsworth.

Coyne, Sarah M., Laura Stockdale, Dean Busby, Bethany Iverson, and David M. Grant. 2011. " 'I luv u:)!': A Descriptive Study of the Media Use of Individuals in Romantic Relationships." *Family Relations*, 60 (April):150–162.

Hall, Jeffrey A. and Nancy K. Baym. 2011. "Calling and Texting (Too Much): Mobile Maintenance Expectations, (Over)dependence, Entrapment, and Friendship Satisfaction." *New Media and Society* 14(2):316–331.

HTC Team. 2013. "Response-Time Expectations in the Internet Age: How Long is Too Long?" High Touch Communications Incorporated, May 27. Retrieved February 12, 2018 (https://blog.htc.ca/2013/05/27/response-time-expectations-in-the-internet-age-how-long-is-too-long/).

Jin, Borae and Namkee Park. 2010. "In-Person Contact Begets Calling and Texting: Interpersonal Motives for Cell Phone Use, Face-to-Face Interaction, and Loneliness." *Cyberpsychology, Behavior, and Social Networking* 13(6):611–618.

Katz, J.E. 2008. "Mainstreamed Mobiles in Daily Life: Perspectives and Prospects." Pp. 443–446 in *Handbook of Mobile Communication Studies*, edited by J.E. Katz. Cambridge, MA: MIT Press.

Lenhart, Amanda. 2012. "Teens, Smartphones and Texting." Pew Research Center, March 19. Retrieved February 17, 2018 (http://www.pewinternet.org/2012/03/19/teens-smartphones-texting/).

Lenhart, Amanda, Aaron Smith and Monica Anderson. 2015. "Teens, Technology and Romantic Relationships: From Flirting to Breaking-Up, Mobile Phones are Woven into Teens' Romantic Lives." Pew Research Center, October 1. Retrieved February 17, 2018 (http://www.pewinternet.org/2015/10/01/teens-technology-and-romantic-relationships/).

Miller-Ott, Aimee E., Lynne Kelly and Robert L. Duran. 2012. "The Effects of Cell Phone Usage Rules on Satisfaction in Romantic Relationships." *Communication Quarterly* 60(1):17–34.

Nielsen Newswire 2011. "New Mobile Obsession US Teens Triple Data Usage." Digital, December 15. Retrieved February 17, 2018 (http://www.nielsen.com/us/en/insights/news/2011/new-mobile-obsession-u-s-teens-triple-data-usage.html).

Prokopets, Elena. nd. "15 Cool and Practical Apps for Couples." *Lifehack*. Retrieved April 27, 2018 (https://www.lifehack.org/articles/technology/15-cool-and-practical-apps-for-couples.html).

Reddit. 2016. "Guys (21+) with Girlfriends: How Often Do You Exchange Texts?" ASKMEN Surveys. Retrieved April 4, 2018 (https://www.Reddit.com/r/AskMen/comments/3hs2rn/guys_21_with_girlfriends_how_often_do_you/).

Reddit. 2015. "Do You Guys Text Your Girlfriends Everyday? And Is It Because You Want to or Is It Because You Feel Obligated?" ASKMEN Surveys. Retrieved April 4, 2018 (https://www.Reddit.com/r/AskMen/comments/248xm7/do_you_guys_text_your_girlfriend_everyday_and_is/).

Schade, Lori Cluff, Johnathan Sandberg, Roy Bean, Dean Busby, and Sarah Coyne. 2013. "Using Technology to Connect in Romantic Relationships: Effects on Attachment, Relationship Satisfaction, and Stability in Emerging Adults." *Journal of Couple & Relationship Therapy* 12(4):314–338.

Yum, Young-Ok and Kazuya Hara. 2005. "Computer-Mediated Relationship Development: A Cross-Cultural Comparison." *Journal of Computer-Mediated Communication* 11(1):133–152.

7

LA FAMILIA IN DIGITAL SPACE AND FACE-TO-FACE

Millennial Latinx Navigating, Customizing, and Reconfiguring Familismo

Cristina Bodinger-deUriarte and Gunnar Valgeirsson

Focus

The increase of Latinx Millennials in the US population, along with fast-paced changes in the use of social media, have converged with considerable social and cultural impact. These 18 to 34-year olds share a cultural tradition of *familismo* that assigns great importance to family, encouraging high levels of interaction and communication among family members. However, like other Millennials, Latinx were raised alongside rapidly expanding digital communication technologies, and they now routinely integrate their *digital presence* and *physical presence* throughout the day. What effect does this routinized co-presence have on how they relate to family members? How do they blend family life with digital life? How aware are they of potential digital encroachment on the time they are sharing the same physical space with family members? One way to approach this question is to identify how the dynamics change (if they do) when one family member no longer has access to digital presence.

The following particularly considers two conditions: (1) *simple digital presence* that involves contact with local family members one may, or may not live with; and (2) *co-presence* and *multi-presence* that involve engaging in online activities or communication technologies that do not include family members, when one is sharing familial physical presence at the same time. Data come from hourly journal entries made by 256 Latinx Millennials who agreed to forego any online or digital communication technology for twelve consecutive waking hours (at a day and start time of their choice). This chapter focuses on the family-related journal entries they made within the context of Latinx overall patterns of digital engagement and their tradition of familismo. The critical concern is the impact that digital presence and multi-presence has on millennial Latinx family relationships.

Why Look at Latinx Millennials as a Group?

It is generally not ideal to use a pan-ethnic category to discuss one particular cultural value. Despite the fact that data are often only collected under these relatively-broad categories, it is important to understand what ethnic subgroups comprise the broader pan-ethnic group and, where possible, note cultural variations if relevant. Latinx is non-gendered and a non-gender-binary version of a pan-ethnic label that describes a particular set of cultural groups of differing national origins (comprised primarily of Spanish-speaking nations) with a composition that varies over time. Flores (2017) provides the following demographic breakdowns. Millennial Latinx living in the US are disproportionately of Mexican ancestry (65%). A very distant second are Caribbean origin groups (16%), more than half of whom are Puerto Rican. Central American origin groups (10%) are almost half Salvadoran. One-third of the small proportion of South American origin Latinx Millennials (6%) are Columbian.

Studies subdividing Latinx into various populations consistently find no notable subgroup variations evident within American Latinx population segments in terms of the cultural values about family under discussion in this chapter. The variation is distinct between Latinx and non-Latinx groups, however "overall, Hispanics exhibit higher levels of familism[o] than non-Hispanics" (Landale, Oropesa, and Bradatan 2006:159). There are two reasons to give particular consideration to Latinx Millennials when exploring this topic:

- Latinx comprise a significant sector of the millennial population
- Latinx use social media more than other millennial groups.

These factors act together so that this population is most likely to demonstrate the broadest number of ways in which physical presence and digital presence unfold within the family, and how they impact family dynamics.

Significant Sector of the Millennial Population

Latinx comprise the second-largest ethnic group in the US (behind Whites), but with lower median ages (twenty-eight for foreign-born and nineteen for US-born Latinx) compared to age forty-three for Whites (Flores 2017). Further, the proportion of this population who are native-born Americans has increased while the proportion of immigrants decreased. Overall, they represent a significant share of the US adult population (15%). Latinx accounted for the largest group of Millennials (21%) next to Whites (56%) in the most recent reports (CNN 2018; Patten 2016). The fact that one in five Millennials are Latinx means that their online behavior has the potential to strongly influence digital use norms and expectations. This potential is increased by the fact that Latinx generally use social media to a greater degree than do non-Latinx. The strongest influence would be expected on the particular sites they frequent more than non-Latinx.

Latinx Use Social Media More than Other Millennial Groups

Just over one in five Millennials identify as Latinx, but does this population use social media any differently than other Millennials? The answer is yes:

- Latinx use of social media sites, *in general,* is higher than other Millennials
- Latinx engage on *particular* social media sites to a greater degree than other Millennials.

Over the last five years, multiple studies focused on millennial social media use have tracked Latinx as the most avid users. By 2013, this group was highlighted as closing the "digital divide" more rapidly than other groups whose socio-economic status affected access to computers and the Internet (Brown, Lopez, and Lopez 2016). It should be noted, however, that gaps in access remain, but are less apparent, due to smartphone technology; Latinx disproportionately access the Internet through only their mobile phones. This was noted by participants who gave up their smartphones for twelve hours:

> New technology innovations shape our daily life. An example is our smartphones that we use on a daily basis to send emails, texts, pictures, etc. or to watch films, stay connected with social networks, or even do homework.
>
> I have discovered how dependent I am on my phone. I rely on it for virtually everything … I not only depend solely on my phone to gain access to the news and my friends, but also to music, something I enjoy quite frequently.

Further, Latinx Millennials are more than twice as likely as other Millennials to report that they use their mobile phones as an expression of who they are. In fact, the "majority of third-generation Hispanics seek more online content that speaks to them as Hispanics" (eMarketer 2017). Nonetheless, by 2013, a greater percentage of Latinx *across age groups* (68%) in comparison with the general population (58%) used social media; while the percentage for Latinx Millennials was much higher (84%) and had reached almost full saturation (94%) by 2016 (Brown, Lopez and Lopez 2016; Lopez, Gonzalez-Barrera, and Patten 2013).

Facebook, with 2.2 billion monthly users, remains, the most popular social media site in the world (Statista 2018). Facebook is also the social media site Latinx use most often; their Twitter use has also increased to bring them even with other millennial groups (Krogstad 2015). Aside from Facebook and Twitter, however, Latinx social media use diverges. For example, Google + is a site few other Millennials use; however, almost half of Latinx Millennials visit Google+ on a daily basis. Far more Latinx get their news (57%) and pop-culture information (31%) through Google+ (Schoon 2016). Less than one-third of non-Latinx Millennials use Snapchat, while over half of Latinx use Snapchat and close to half use Instagram on a daily basis (Schoon 2016).

Familismo

Latinx use the term *familismo* to indicate the complex importance of family. Familismo emphasizes "family as the major source of one's identity and protection against the hardships of life. This sense of family belonging is intense … The family model is an extended one; grandparents, aunts, cousins …" (Carteret 2011). Family studies in the early 1990s identified three dimensions of familismo: family structure, attitudes, and behaviors (Landale, Oropesa, and Bradatan 2006). Its cultural importance is indicated by the fact that there is a specific label for this multi-dimensional set of values embracing attitudes, behavior, and an extended family structure. A brief discussion of these dimensions precedes a discussion of how doing familismo has been impacted by digital life.

Family Structure

Elements of family structure include the presence or absence of nuclear and extended kin, typical marital age, childbearing age expectations, and family size. Some scholars predicted that US native-born Latinx would assimilate away from the overall values of familismo (Broadie et al. 2002). Some diminishing of familismo has been observed; however, this has primarily been in three aspects of family structure: lower fertility rates and later ages for marriage and childbirth. Otherwise, the familismo structural elements persist across all subgroups of Latinx (Landale, Oropesa, and Bradatan 2006:152) particularly in terms of living with family where "households of Hispanics are slightly larger and much more likely to be extended than those of non-Hispanic whites" (Ibid:168). In particular, when all Latinx subgroups are compared to non-Latinx US populations, the "most striking" contrast is that Latinx elderly "are considerably more likely to live with other relatives and less likely to live alone" (Landale, Oropesa, and Bradatan 2006:155). Where do Millennials fit into family structure?

Millennials Living with Parents

According to a Pew Research Center study, "2014 appears to be a milestone … For the first time since 1880, young adults are more likely to be living with a parent than they are to be living with a romantic partner in their own household" (Fry 2016:10). This is true across demographic groups, includes college undergraduates, and is not accounted for entirely by socio-economic class. For Latinx Millennials, 2011 was the milestone leading to the current figures of less than a third living with a significant other, while higher percentages of this 18–34-year-old group of Latino men (38%) and Latina women (35%) live with parents (Fry 2016:8). Latinx Millennials show this in their journals through their many references to parents, siblings and grandparents in the household. These multi-generational family households reflect the structural component of familismo, but do not guarantee that attitude and behavior components also persist.

Family Attitudes

The attitudes associated with familismo not only emphasize the overall importance of the family, but also include a specific set of values. Solidarity among family members is critical, along with the closely-related value of loyalty. There is also an expectation of "give and take" that stresses the importance of reciprocation (Landale, Oropesa, and Bradatan 2006). Both foreign-born and American-born Latinx adolescents presented more attitudes reflecting parental respect, and a higher sense of reciprocity, than did non-Latinx Whites (Fuligni, Tseng and Lam 1999). Although studies note some attitudinal changes among later generations of second- and third-generation Latinx, these cluster around generational shifts towards more progressive ideologies. This could be interpreted as an erosion of filial solidarity. However, recent work (Leonard and Lugo-Lugo 2015; Villa 2016) agrees with earlier study findings that although "those whose ancestors have been in the United States for multiple generations express weaker emphasis on family, they are still more likely than non-Latinos to agree with views that underscore the importance of family" (Broadie et al. 2002:37). Such attitudes definitely surface in the journal entries.

Family Behaviors

The behavioral dimension of familismo includes family role obligations, sharing economic resources, mutual assistance and social support, and frequent contact among family members (Landale, Oropesa, and Bradatan 2006). The persistence of these behavioral components have been repeatedly verified in studies of Latinx families for more than two decades (Kanellos 1994; Leonard and Lugo-Lugo 2015). For example, Kim and McKenry (1998) found that Latinx (whether aggregated or disaggregated by nationality of origin) tended to socialize more frequently with relatives than did other racial and ethnic groups. Again the journal entries included numerous references to time spent with cousins, nephews, nieces, grandparents, parents, siblings, and the parents and other relatives of romantic partners as well.

Fuligni, Tseng, and Lam reported that both immigrant and native US Latinx show more filial obligation and a willingness to support aging parents than non-Latinx Whites (1999). A review of the 2004 National Survey of Families and Households assessed "numerous questions that tap normative beliefs about the obligations of parents to support their adult children and the obligations of adult children to support aging parents" and found that members of all Latinx groups were "more likely than non-Hispanic whites to recognize both parental and filial obligations" (Landale, Oropesa, and Bradatan 2006:169). Similarly, Villa reports that "a recurring theme among Hispanic Millennials is the central role that family plays in their lives" (2016).

Interfacing Family

Millennials grew up in a context that involved *multi-presence*—that is, sharing physical space while more than one person is simultaneously engaging both in *physical presence* and separate, unshared, *digital presence*. This has been the case from the time they were old enough to make conscious choices about how to spend their time with family and friends. Family-interaction choices include:

- *Simple digital presence:* Maintain connectivity when not together through texting, email, and other online/smartphone activity (even if sharing a household).
- *Multi-present connectivity:* Directly interacting with family members who are physically present, while *also* integrating online material into these interactions (through sharing digital photos, entertainment, information, social media, or other online material with one another).
- *Co-presence or Multi-presence*: Directly interacting with family members who are physically present, while engaging in unshared online activity that does not include them. Co-presence refers to this behavior by a single member, while multi-presence indicates that more than one person is engaging in co-presence. In other words, that co-presence is enacted by more than one family member at the same time.
- *Physical presence:* Focus on in-person interactions and forego digital presence while with family.

Data indicate that many millennial Latinx have not experienced significant spans of time interacting with family by sharing physical-only presence. Consequently, before participating in the twelve hours of abstinence, most participants did not have an adequate frame of reference from which to judge whether their digital life had an impact on family dynamics. Ultimately, they did not have the perspective to ascertain whether different choices around digital activity would better align with their views of family affiliation.

Family Ties in Digital Space

As early as 2008, about one-quarter (23%) of survey respondents reported that the Internet had improved family communications (Wellman et al. 2008:iv). Latinx Millennials use some degree of their digital presence to "do familismo," and report this as a conscious choice.

> Not only do they remain close to family, they adopt social media as a way of staying connected with family. For instance, when they were asked about the different reasons that they use Snapchat, 51% reported that they did so in order to stay connected with family. This is roughly 20 percentage points more likely than any other Millennial cohort (Villa 2016).

Non-Millennials and non-Latinx also reach across digital space to touch base with family members who are local. In fact, as of 2012, almost half of the general population "(42%) of parents call their children from cell phones at least once per day" (Hertlein 2012:381). This rate is higher for Latinx parents who use social media like Snapchat, as well as texting to maintain daily family connectivity. Even if only working parents are considered, almost half text from work (47%) or use social media while at work (42%) for non-work-related reasons (Common Sense Media 2016).

Journal entries support earlier findings that feeling satisfied about using one's cell phone is related to feelings of connectivity and relationship satisfaction with one's mother and father; frequency of cell phone interaction was also positively associated with feeling close to one's mother. (Miller-Ott, Kelly, and Duran 2014:313). The twelve hours offline experience underscored this aspect of family connectivity. Both millennial parents and younger adult Millennials frequently referred to missing their digital connectivity with family members.

> I began to miss my phone [to text] because I wanted to make sure my daughter was fine ... Not having my phone did make me miss my daughter more.
>
> Me and my mom always keep in touch, she always wants to make sure I am safe and doing good .... we always check up on each other. We have a pretty close relationship and worry about each other.
>
> The real problem I noticed was that I didn't feel connected with my friends and family. Not having access to certain communication apps really had toll on me.

Although texting has become more prevalent than email, the current study also supports the dynamic identified by Eppler and Walker (2004); they found that siblings who emailed one another experienced greater levels of commitment and trust. As one participant wrote:

> Not being able to just send a text message to my sister and family at home, made me feel further from home and disconnected from the world.

Journal entries indicated that the family members they live with, particularly mothers and siblings, remain an important daily touchstone during temporary absences from home.

According to more recent studies, daily texting with local family members not only includes "general conversation," it is also used in pragmatic ways that reflect familismo values of reciprocity and affiliation. Texts convey information about family matters, plan family activities, and address other family concerns (Crosswhite, Rice, and Asay 2014). It appears that simple digital presence has generally positive outcomes for familismo, when it is specifically used to build connectivity during times when family members are not also sharing physical space. However, what if they are also physically present?

Multi-Present Connectivity the Family Way

Journal data also address how this population interfaces family through shared multi-presence, that is, through multi-present connectivity. The digital availability of many forms of entertainment can be seen as an opportunity to connect. In fact, almost a decade ago, a major survey found that families were more likely to interact *with one another* online when individual members had their own computers (Wellman et al. 2008). This seems to be supported by a number of journal entries that decried the loss of shared digital presence (multi-present connectivity) as a family activity. This was particularly true in terms of viewing YouTube videos and computer and tablet access to Netflix movies and shows.

> My mom felt lonely because we would watch funny movies … and she didn't get to share the enjoyment with me.
>
> That made me sad because I realized watching [programs] is sometimes how we "bonded." I would have been sitting there with her talking and watching.
>
> We usually have breakfast and watch funny videos or take funny pictures with Snapchat.

Families who access digitally-provided videos, films, and other programming together might be seen as parallel to family television viewing in an earlier era. What families share in digital space today, may not be essentially different from what they shared before the digital age. Even if family togetherness shifted to digital platforms, many of these are analogs. In other words, they are similar in nature to pre-Internet activities, but are "translated" into parallel digital activities—the way playing Words with Friends online, while in the same room, approximates a board game type of experience. A number of years ago, a major study from Pew Research Center claimed that:

> Although some commentators have expressed fears that technology pulls families apart, this survey finds that couples use their phones to connect and coordinate their lives, especially if they have children at home … when they return home, they often have shared moments of exploration and entertainment on the internet (Wellman et al. 2008:ii).

Outcomes of the current study are less positive.

Three questions are key in looking at the family dynamics in the multi-present connectivity context: (1) Are family members motivated by family bonds to connect through sharing digital activity? (2) Are individuals motivated to maintain their own digital presence, and only sporadically connect with family members as they do so? (3) Is the intention to do the former, while behavioral choices look more like the latter?

The Gap Between Intention and Interaction

Familismo perspectives align with putting family affiliation ahead of digital presence. More individualist perspectives, where the family is not as central, would put one's own digital presence ahead of connecting with family members. Latinx claim a primary reason they engage in social media is to connect with family. Journal entries, however, overwhelmingly reveal a relatively small degree of multi-present connectivity. Rather, family members are much more likely to simply be multi-present—that is to be engaged in digital activities that do not include the family members with whom they are sharing physical space.

The current study clearly supported claims of positive family connectivity through simple digital presence when *temporarily away* from the shared household. Although mentioned, relatively few journal entries referred to multi-present connectivity where family members simultaneously share both digital and physical presence with one another.

Being Co-present and Multi-Present at Home

The need to engage in texting, social media, and other Internet activities with non-family leads to losses in interaction and conversation with family, and low participation in much of daily family life, despite sharing the same physical space. Ironically, the picture that emerges is almost one in which the family demonstrates higher connectivity when absent, than when present. This is reflected in Latinx journal entries from both younger millennial adult children, and from young millennial parents:

> I sit down with my mother in the kitchen for a little while. I have been so busy or distracted by my phone that I have not had the time to sit and talk with my mother the entire week [lives at home with mom].
>
> I feel good about not checking my phone. I feel the baby is bonding with me more this morning while feeding. I can focus all my attention towards her. I am used to feeding the baby with one hand and having my smartphone in the other. I'm actually ashamed of admitting to that. It doesn't make me sound like a good mother.
>
> Most of the time my family says I do not pay attention to them, which is true. They can be talking to me and I won't listen to a thing they say because I am preoccupied with whatever is on my phone. Unplugging made me direct my attention to what is actually going on around me, opposed to what is on a screen.

Over the last decade, multi-present-connectivity at home has shifted to more self-oriented multi-presence. The contrast is clear in view of how participant journal entries in the current study compare to earlier findings from Pew Research Center that describe higher connectivity at the home front.

> Yes, Facebook could be a way to keep in touch with family members … but most of the time I was looking at what my friends were posting rather than connecting with family members.
>
> I sat in my living room still reading my book and analyzing my family. I realized that even though we were all sitting together in the living room, they were all consumed by their phones, and basically in their own little world … I would of probably been doing the same as them.

Contrast this to the picture from a ten-year-old study:

> Although families often have the means to retreat to their technological "neutral corners" and engage in screen time in isolation, many go online with others. Some 52% of Internet users who live with a spouse and one or more children go online with another person at least a few times a week. Another 34% of such families

> have shared screen moments at least occasionally (Wellman et al. 2008:iii).

Connectivity rates described as a "few times a week" or "at least occasionally" may have been appropriate a decade ago, in a lower use era. However, with skyrocketing use rates over the last decade that picture of connectivity would no longer suffice. As of 2011, Millennials in their twenties sent between 100 to 134 text messages per day in addition to their social media use, online gaming, and their digital video, film, and program consumption (Nielson Newswire 2011). This represents a threefold increase a mere three years after Wellman and his colleagues were writing. Aside from text messaging, Latinx *parents* of tweens and teens engage in nine hours of digital screen time per day that is not work-related (Common Sense Media 2016). However, despite spending so much more time in digital space, multi-present connectivity rates remain relatively low. Consider the following example:

> I was so irritated because I was talking to my mom and she literally ignored me for five minutes because she was "checking her Facebook." I felt mad and just irritated because I never really paid attention to that. Instead, I got up from where I was sitting picked up the baby and took her outside for a bit just so I can cool off. I have never been ignored in such way and I just started thinking that what if I do the exact same thing to others without noticing.

Dinner with the Family

Shared meals are the iconic symbol of family life. However, family meals were an early site of multi-presence. In their otherwise optimistic study, Wellman and his colleagues discovered:

> Those with multiple communication devices are somewhat less likely to eat dinner with other household members and somewhat less likely to report high levels of satisfaction with their family and leisure time than are families with lower levels of technology (2008:iii).

Ten years later, journal entries in the current study support this observation time and time again. This is the clearest evidence of multi-presence displacing the dynamics associated with familismo.

> Decided to cook dinner for my family with the help of my mother. I would otherwise have been watching a documentary online, or on YouTube.

> I got the opportunity to cook for my nephews and my son instead of lying in bed watching Netflix. Since my family members usually see me on my phone I got the opportunity to create an interaction while enjoying breakfast.
>
> Instead of being glued to my phone while we ate, I was able to help my mom with setting up the table and helping prepare for breakfast. Not only that, but I also helped clean up such as washing the dishes.
>
> My young boys seemed bothered and my significant other was skeptical at first … We instead [of going online] had a conversation of how our day was going to be planned out. I enjoyed spending quality time at the breakfast table with my family.

For many, habitual distancing behaviors in multi-presence seem unintentional. It took the change in a family member's access to digital space to illuminate it.

> My family and I just had dinner together but this time we did something very different. In attempts to help me with my [offline status] my mom ordered everyone to set aside their phones at the dinner table, something that is very rare for my family. This experience was something I was not really use too but I really enjoyed it. For once we had actual bonding time where we talked about each other's day rather than eating in silence consumed by our phones. This made me realize that technology takes over us in so many ways and sometimes we fail to see that.

How Other Family Members Reacted

Two-thirds of those foregoing digital presence (66%) chose to increase the time they spent "in-person." How did this extra time affect their family members?

> My mom said that she was happy to see that my hands were not stuck to my cell phone. She and I spent the hour talking… My parents are loving this "no-technology day." They keep saying that they are going to take my phone away from me more often.
>
> My dad seems to be enjoying the fact that I am actually talking to him instead of on my phone.
>
> I like seeing my daughter happy. She seems to be enjoying outdoor play. I am starting to feel guilty for always staying indoors and not taking my daughter outside more often.
>
> I was having a great time with the family. I felt that stepping away from technology actually gave me the opportunity to spend

> quality time with my family. The kids actually had their uncle participating in the games and had as much fun as I did by me being part of the activities.
>
> My brother is ten years old, so he is at a stage that just loves playing and bothering me like crazy … . It is rare that I play with him. Today is his lucky day, I get to play with him the whole day if he wished … I would [usually] be playing with my phone or at work doing something to distract me, but instead I'm playing with my brother … . My brother is happy that I have not stopped playing with him … . My mom got home to see me playing with my brother. She was happy to see me like this because it is rare.
>
> I saw my sister reading a book so I started asking her what she was reading. She was more than happy to tell me what the book was about. I realized that communicating with my younger sister is very important. It's something I hadn't noticed because I get distracted with my social life. I was no longer viewing my "time left with no phone," and just began living the moment.

Not all of it is positive, however. Digital presence is ingrained into daily routines and may not always be readily foregone, despite another family member's digital abstinence.

> As I sat down with my family to eat it felt weird not being on social media like everyone else at the table. I noticed I was the one trying to make a conversation but they weren't really collaborating with me. I was starting to feel frustrated.

Post-millennial siblings sometimes also put their own digital presence ahead of connecting with family members. This could be anticipated, however, as they may well never have experienced anything different. A participant trying to bond with a sibling wrote:

> My younger sister is always on her phone and she is always around me and seemed so caught up and wouldn't really pay attention to me when I would talk to her because she was more interested in what her friends on "twitter" were putting.
>
> My youngest sister … kept on stopping our conversations because she would rub it in my face that she was able to use her phone and check all her social media outlets.
>
> I had to eat by my lonely self in my room because everyone wanted to watch a movie … At this point not only was I kept from being part of the online world, I was being cast out by my own siblings. Even my six year old nephew tossed me out. I asked him if he wanted to keep me company in my room and he said

"No, I don't want to talk to you because you can't watch [the online film].

Balancing Presences and Recognizing Priorities

This digital abstinence had a profound effect on many of the participants. Recognizing multi-presence as a distraction was one of the most common participant insights:

> I usually spend as much time with my family as I did today. The only difference was that this time we didn't really have much distraction while we shared this time. I listened to my sister's conversation … and we all were not being distracted. It was nice to know that we can have a family moment. I can say we had a real family moment.
>
> I'm enjoying this experience. I had not realized how much time social media takes away from the family that's in front of you. I have been able to focus on my two daughters a lot more. I also feel guilty for being so involved in my phone and social media.

Millennials have commonly seen multi-presence as simply multi-tasking without cost. The change in this perspective proved to be transformative for many participants:

> As soon as I hopped on the bus with my mom, she sat right next to me and took out her phone and headphones and tuned me out … Her response to my facial expression was, "Well you do that when you are on the bus with me." This just opened my eyes more to that in fact I … could be hurting other people's feelings … she stepped off the bus saying, "Call me! Or text me!" once again making fun of me and making me realize more and more that my phone was taking up some of my family bonding time.
>
> For my last hour of, this project, me and my boyfriend spent an hour playing *Uno* and *Trouble* with his little sisters. They enjoyed it as well because we hardly do things like that. It made me happy to see them excited and I then realized that I wanted to keep some of these things going on after this project. I am thankful for what it taught me.
>
> [Generally] I would waste more time on all of them [social media sites] than using my time to do things around the house and spend more time with my daughter. I have actually decided that Sundays will be the day where I will limit my technology use and will make it a day spent with my daughter, family and friends.

> Observing my family made me realize that we waste so much of our time being consumed in the media and technology that sometimes we let the good things in life pass us by. Instead of my family being on their phones, they could of been engaging with each other and enjoying quality time.
>
> Undoubtedly, the millennial era has allowed more accessibility to technology, but finding the balance where accessibility of technology does not dictate my life is now a challenge for me. Because I am fully aware of this, I can take the appropriate steps to appreciate my loved ones more, slowly change my overall urges, and hopefully, if successful, have a life with meaningful substance.

In fact, a significant number of participants (42%) specifically mentioned being more aware of the improved quality of in-person interactions. Just over a third (36%) said they believed it would be important to remain aware of the balance between their digital and physical presence, in general. Many also mentioned recognizing the need to be more fully engaged with family in physical presence (29%).

Last Word

The current study findings may reflect several factors acting together to undermine familismo. First, the improvement of smartphone technology has closed much of the digital divide so that family members each have their own Internet access devices. This enables Latinx Millennials to further raise their already distinctly high levels of digital activity. Escalating expectations around digital texting and social media participation and response times create a digital time demand that inevitably encroaches on the centrality of engaged family time while sharing physical presence. However, perhaps the most significant factor that surfaced across these 256 Latinx journals is the overwhelming indication that such encroachment was unintended, and may well have remained unnoticed without the perspective gained by the relatively simple offline experience.

The power of the draw into digital space frequently overwhelms family members who have every intention of affiliating regularly and meaningfully. Having grown up in the digital age, Millennials are often too inexperienced in simple physical presence to recognize the increasing imbalance between digital and physical presence in their familial interactions. Consequently, participants frequently reported that this relatively brief digital abstinence experience resulted in a significant perspective change, and a stronger awareness from which to reconsider choices around balancing their digital presence and physical presence. This is a sobering realization given the unlikelihood of widespread replication.

The most important take-away from this chapter seems to be that the traditional Latinx emphasis on family-centric values remains strong among Millennials, even as they unwittingly undermine the very connectivity that supports familismo, by being overly multi-present. A participating Latinx Millennial father puts it well:

> The essence of not having technology run my life allowed me to participate in my family activities which I have been absent from for many years … Instagram has caused me to distance myself from the family. I don't have to continue being controlled by technology … This exercise made me realize that life is more beautiful and for that reason we should limit the usage of technology and I have made a promise to myself that when around family I will put my phone away … Sometimes people lose track of what is real important in life.

References

Broadie, Mollyann, Annie Steffenson, Jaime Valdez, Rebecca Levin, and Roberto Suro. 2002. "2002 National Survey Of Latinos." Pew Hispanic Center/Kaiser Family Foundation, December. Retrieved February 20, 2017 (http://www.pewhispanic.org/2002/12/17/pew-hispanic-centerkaiser-family-foundation-2002-national-survey-of-latinos/).

Brown, Anna, Gustavo Lopez and Mark Hugo Lopez. 2016. "Digital Divide Narrows for Latinos as More Spanish Speakers and Immigrants go Online." Pew Research Center: Hispanic Trends, July 20. Retrieved March 31, 2018. (http://www.pewhispanic.org/2016/07/20/3-hispanics-and-mobile-access-to-the-internet/).

Carteret, Marcia. 2011. "Cultural Values of Latino Patients and Families." *Dimensions of Culture*, March 15. Retrieved March 28, 2018 (http://www.dimensionsofculture.com/2011/03/cultural-values-of-latino-patients-and-families/).

CNN. 2018. "Millennial Generation Bigger, More Diverse than Boomers." *CNN Money*, January. Retrieved March 31, 2018 (http://money.cnn.com/interactive/economy/diversity-millennials-boomers/).

Common Sense Media. 2016. "Hispanic Parenting in the Digital Age." Common Sense Media Census. Retrieved April 20, 2016 (https://www.commonsensemedia.org/research/the-common-sense-census-plugged-in-parents-of-tweens-and-teens-2016).

Crosswhite, Jennifer M., Denise Rice, and Sylvia Asay. 2014. "Texting among United States Young Adults: An Exploratory Study on Texting and its Use within Families." *The Social Science Journal* 51(1):70–78.

eMarketer. 2017. "US Hispanics and Digital Media Usage: How They Differ from Non-Hispanics and from One Another." eMarketer Report, June 13. Retrieved March 31, 2018 (https://www.emarketer.com/Report/US-Hispanics-Digital-Usage-How-They-Differ-Non-Hispanicsand-One-Another/2002074).

Eppler, Christie. and Erin K. Walker. 2004. "Electronic Ecology: An Ecosystemic Exploration of Sisters Communicating Via Email." *Journal of Feminist Family Therapy*. 16(1):19–37.

Flores, Antonio. 2017. "How the US Hispanic Population is Changing." Pew Research Center, September 18. Retrieved March 29, 2018 (http://www.pewresearch.org/fact-tank/2017/09/18/how-the-u-s-hispanic-population-is-changing/).

Fry, Richard. 2016. "For First time in Modern Era, Living with Parents Edges out Other Living Arrangements for 18–34 Year Olds." Pew Research Center, May 24. Retrieved September 10, 2018 (http://www.pewsocialtrends.org/2016/05/24/for-first-time-in-modern-era-living-with-parents-edges-out-other-living-arrangements-for-18-to-34-year-olds/).

Fuligni, Andrew J., Vivian Tseng, and May Lam. 1999 "Attitudes toward Family Obligations among American Adolescents with Asian, Latin American, and European Backgrounds." *Child Development* 70(4):1030–1044.

Hertlein, Katherine M. 2012. "Digital Dwelling: Technology in Couple and Family Relationships." *Family Relations: Interdisciplinary Journal of Applied Family Studies* 61(3):374–387.

Kanellos, Nicolas. 1994. "The Culture of Latinos in the United States" Pp. 15–38 in *Handbook of Hispanic Cultures in the United States: Anthropology*, edited by Thomas Weaver and Claudio Esteva-Fabregat. Houston: Arte Público Press.

Kim, Hyoun K. and Patrick C. McKenry. 1998. "Social Networks and Support: A Comparison of African Americans, Asian Americans, Caucasians, and Hispanics." *Journal of Comparative Family Studies* 29(2):313–334

Krogstad, Jens Manuel. 2015. "Social Media Preferences Vary by Race and Ethnicity." FacTank News in the Numbers: Pew Research Center, February 3. Retrieved March 28, 2018 (http://www.pewresearch.org/fact-tank/2015/02/03/social-media-preferences-vary-by-race-and-ethnicity/).

Landale, Nancy S., R. Salvadore Oropesa, and Cristina Bradatan. 2006. "Hispanic Families in the United States: Family Structure and Process in an Era of Social Change." Pp. 138–178 in *Hispanics and the Future of America*, edited by Marta Tienda and Faith Mitchell. Washington DC: National Academies Press.

Leonard, David J. and Carmen R. Lugo-Lugo, eds. 2015. *Latino History and Culture: An Encyclopedia*. New York: Routledge.

Lopez, Mark Hugo, Ana Gonzalez-Barrera and Eileen Patten. 2013. "Closing the Digital Divide: Latinos and Technology Adoption." Pew Research Center: Hispanic Trends, March 7. Retrieved March 29 2018 (http://www.pewhispanic.org/2013/03/07/vi-social-networking/).

Miller-Ott, Aimee E., Lynne Kelly, and Robert L. Duran. 2014. "Cell Phone Usage Expectations, Closeness, and Relationship Satisfaction between Parents and Their Emerging Adults in College." *Emerging Adulthood* 2(4):313–323.

Nielsen Newswire. 2011. "New Mobile Obsession US Teens Triple Data Usage." Digital, December 15. Accessed February 17, 2018 (http://www.nielsen.com/us/en/insights/news/2011/new-mobile-obsession-u-s-teens-triple-data-usage.html).

Patten, Eileen. 2016. "The Nation's Latino Population Is Defined by Its Youth." Pew Research Center: Hispanic Trends, April 20. Retrieved March 31, 2018 (http://www.pewhispanic.org/2016/04/20/the-nations-latino-population-is-defined-by-its-youth/).

Schoon, Robert. 2016. "How Latino Millennials Compare to Peers in Mobile, Social Media Use – New Report." *Latin Post*, May 5. Retrieved March 29, 2018 (http://www.latinpost.com/articles/120212/20160505/how-latino-millennials-compare-to-peers-in-mobile-social-media-use-report.htm).

Statista. 2018. "Most Famous Social Network Sites Worldwide as of January 2018, Ranked by Number of Active Users (in Millions)." The Statistics Portal. Retrieved March 31, 2018 (https://www.statista.com/statistics/272014/global-social-networks-ranked-by-number-of-users/).

Villa, Jose. 2016. "The Social And Mobile World Of Hispanic Millennials." *MediaPost*, May 26. Retrieved March 28, 2018 (https://www.mediapost.com/publications/article/276765/the-social-and-mobile-world-of-hispanic-millennial.html).

Wellman, Barry, Aaron Smith, Amy Wells and Tracy Kennedy. 2008. "Networked Families." Pew Research Center, October 9. Retrieved April 20, 2018 (http://www.pewinternet.org/2008/10/19/networked-families/).

8

IS UNMEDIATED MORE?

When Physical Presence Does Not Equate to Digital Presence

Berge Apardian and Cristina Bodinger-deUriarte

Introduction

Millennials have grown up within a social context in which people simultaneously exist in two different environments, the physical and the digital. Having a *physical presence* differs from that of a *digital presence*. The digital presence is facilitated by digital platforms in which people can create pseudo identities and communicate through a medium. When entering this digital space, individuals take part in an environment in which disembodied communication takes place. This is facilitated through their own self-created profiles, often accompanied by status posts demonstrating current mental or emotional status. While in digital space, individuals disengage from their physical presence, generally momentarily or sporadically, yet attempt to also be physically present with others. This chapter presents the argument that this sort of co-presence has a tremendous impact on the quality of communication and interaction that people experience. The current study was based on 420 Millennial college students giving up all forms of digital activity for twelve consecutive waking hours, while making hourly journal entries about their "unplugged" experiences. Participants chose their start time and also the date (within a several-week window). These journal entries support the idea that people often engage in their digital presence at the expense of the quality of their interactions with those sharing their physical presence.

In the Company of Multi-Presence

The quality of physical communication becomes (at a minimum) halved, when an individual is trying to be physically present with companions, and digitally present elsewhere (co-present). Communicative quality is further compromised when *multiple* individuals are *simultaneously* co-present, resulting in a *multi-presence* among them. The journal narratives give the impression of continuous engagement with digital devices, and near constant distraction by digital activities. Whether they are having dinner with family or friends, running in the park, reading to their children, or doing their homework, they are typically also on their devices. There are different ways of doing multi-presence, some of which are positive, but the problematic aspect is when there is a loss of connectivity with those present. For example:

> I went to the table and ate my dinner with my mom and sister. And we talked, we caught up on what's been going on. My mom explained to my sister how great her day was with me because I didn't have my phone. My sister thought this assignment was so funny, and she, too, was happy I didn't have my phone.
>
> I was able to read to my daughter without having any distractions and not read while multitasking. She was able to have my full attention and this was definitely beneficial to my daughter because if my phone had gone off I would have answered it, and if someone was on TV that interested me, I would have stopped to watch it.

Young millennial parents routinely feel a need to be digitally present, even when their children would greatly benefit from uninterrupted parental attention when they are together. This sort of routine multi-presence creates a generation of individuals who feel the constant need to multitask, and who seem impelled to digitally communicate with one set of people, even when they share physical presence with another person or group of people. Participants often wrote about how being fully physically present enabled them to recognize that their previous co-presence decreased the quality of the communication they had with others.

Participant journal entries frequently included reflections on the relative quality of conversation when both individuals are fully physically present.

> I listened to my mother better rather than being focused on my phone. My mom and I also had a great conversation during this trial.
>
> I have realized after doing this experiment that it is more rewarding to actually talk to someone face-to-face than just reading what they are saying, and trying to picture how they would be having a conversation with you in person.

However, co-presence is currently accepted as normative because Millennials brought-up in this type of environment do not necessarily know what it is like to forgo multi-presence for significant periods of time. This highly effects their face-to-face communication. The impact of co-presence can be clearly visible to those who are fully physically present—responsive and expressive eye-contact and body language are diminished or supplanted by meaningless glances-up from a device. Conversational reactions are suspended, delayed, or supplanted by a few chuckles while posting or texting with someone else. This is made worse when digital discussions too frequently replace those in a shared physical environment. Current studies have shown that even the basic skills necessarily for effective communication, like the recognition of interpersonal cues, can be affected when online communication is over-used and people are "not seeing facial expressions or using nonverbal communications" because they "limit the opportunity to practice in-person conversations that are crucial" (University of North Carolina News 2016).

Often an escape, mobile devices can easily pull people into their digital presence. This can deeply impact communication skills, and the simplicity of interacting with others, at worst, it can contribute to social anxiety. The more frequently someone falls back on their mobile device as a shield from others, the more likely they are to feel hesitancy in meeting new people or more generalized anxiety about taking part in social events. On the other hand, going to a party without a smartphone may even compel a generally introverted person to interact with others.

> I don't know what to talk about and will turn to fidgeting with my phone. I felt awkward and a little lost. I didn't know what to talk about and I was pretty quiet at first but I warmed up quicker than I would normally, and was soon talking to many people. Instead of playing with my phone, I was more social and I enjoyed myself, and meeting new people. I was happy.

Spending Time without Distractions

Every free moment is a moment people can look to their devices. Millennials have integrated digital activity into any time spent waiting for something, even when they are accompanied. Journal entries show this type of multi-presence to be completely routine.

> I have realized that technology has ruined quality time during all times of the day. My breakfast, lunch, and dinner with family and friends with no usage of cell phones, enabled us to bond together, talking about old memories, laughing; something we as a family have not done in a while … this project has enabled me to reconsider the usage I do on my phone when around family and friends.

> Normally, while waiting for food to come out, we would be playing with cell phones, checking out videos and Facebook, or playing video games. However, instead, we talked. I felt that it was such a long time ago I had a full conversation with my husband. It was fun. Our full attentions were on ourselves. These distractions have been disrupting our relationship.

Husbands and wives using the time to communicate with one another, rather than turn to separate digital spaces, is more likely to occur when people have a better understanding of the importance of fully-engaged communication, and the need to "unplug" for a while.

> I spend time with my husband eating dinner, and talking to him about our day. Normally we go to our computers and do our own thing.

Dealing with one another directly has the potential to save relationships, because it takes away a distraction, and removes an opportunity to escape an uneasy discussion. For example, one participant wrote:

> I had been neglecting our relationship for a while. I started to discuss our relationship with my husband. At first, he did not want to talk, so I started to open up and talk honestly. He started to open up too. We realized that there were many issues that had been built up and never discussed. If there was full access to the Internet, I would have been checking my emails instead.

Being physically present provides opportunities for people to discuss issues in general. Further, relationship satisfaction can be eroded when intimates believe their partners are using digital presence as a method of avoidance (Blatchford 2017). Physical presence decreases opportunities to evade what may be a seemingly unpleasant, yet meaningful, discussion. In addition to the relationship threat from conflict avoidance, a recent study shows that those who overuse social media to communicate may also fail to develop adequate "conflict management skills" (University of North Carolina News 2016).

Deferring to the Digital

Picture this scenario: a family of four has the television on; the mother and father are on their laptops; the brother and sister are each on their own tablets; they find themselves physically together, but are not engaging with one another. They are in the same room, but there is no connectivity with one another. This is a reminder of how digital presence can replace active physical presence. When one is used to replacing active physical presence, even music can displace interaction and no longer becomes "background music." This was frequently mentioned in the journals.

> I drop my sister at work, which is a 40-minute drive. I usually listen to my ipod in the car, but I just turned it off and had an actual conversation with my sister. It felt great because we really never talk even though we share the same room.

These sisters who share a room without really talking, do not dislike one another's company; they are simply unaware of the lack of engagement when they are together because they have become used to routinely shifting their attention to their devices.

As mentioned, "wait time" is a typical trigger for multi-presence. When waiting to arrive at a destination, or for food to arrive at a restaurant, or while standing in line for ice cream or a movie, or when taking a break from work, or walking between college classes, it is common for Millennials to automatically gravitate into digital presence. Imagine the amount of conversation that is lost.

One explanation for this is that being disconnected from digital space has come to be equated with "being alone," even when among others. In fact, a significant number of journal entries mentioned that forgoing digital presence made them feel "disconnected" (28%) and lonely (23%) despite the fact that they were, in many cases, writing these things while at home with family members or out with friends. Being digitally present and reacting to comments with clicks and emojis creates a feeling of community that satisfies a sense of connection. Receiving comments, emojis, "likes," and other digital reactions sparks a feeling of approval within the person, and indicates that engaging in the same types of responses is favorable behavior. This acts to secure the reproduction and continuity of sharing people's experiences with clicks and posts. Studies have shown that the very fact we have the ability to engage with one another in digital space, creates an escalating expectation that we will do so, and that the more we participate, the greater the expectation for *additional* participation. This can lead to a constant need to share, describe, and receive positive feedback, about experiences, despite being in the physical company of others, just to satisfy expectations (Hall and Baym 2011). As one participant wrote:

> I am always texting my boyfriend when we are not together, or even friends about how our day went. We hardly go grab coffees or go over to each other's houses if it's not for the purpose of going out. But, when we do go out we are still trying to capture every moment on our phone by taking pictures and video.

It is almost a reflex to regularly glance at cell phones to see if there is a message, or to check the time. Doing so can also cause people to automatically and habitually dive into digital presence. A simple glance at a phone to check the time, may turn into a fifteen-minute status update. In fact, when checking ones texts, new phone apps trigger invitations to play particular games, or post on particular sites popular with the owner of that phone. This entices prolonged engagement when the original intention was just to quickly check something. The following journal entry provides an example:

> It made me realize that I need to buy a watch so I won't have to look at my phone for the time. Every time I looked at my phone, I would text and play Words with Friends … It made me appreciate conversations in person more than through phones and emails. I am on my phone frequently when I really don't have to.

Studies have shown this to be typical of millennial behavior. When a text message pings, or a tweet or Facebook message posts, users typically take twenty– twenty-five minutes away from whatever they had been doing. However for almost one-third of those checking in (30%), it takes *two hours* to resume a former activity (Gasser and Palfrey 2009). Knowing that one's device is always within reach, makes it easy to adopt the habit of constantly monitoring social media.

Text or Talk? That is the Question

Multi-presence may seem to be just a means to an end, but it often replaces physical communication, taking away the desire to physically meet. The thirst to interact is satisfied. A study by the Center for the Digital Future (2012) argues that digital presence increasingly usurps physical presence among family members sharing a household. In 2002, the impact was already becoming evident with a relatively small, but significant number of people reporting they were spending less time with family members who shared their household (8%); by 2011, with the advent of texting, a third (34%) reported spending less face-to-face time with physically-present family members. This results in situations in which we text those in the same household we share with them, despite their proximity. As one participant put it:

> It seems that more and more people are moving away from face-to-face interactions, but instead moving towards a more technology-driven way of life.
>
> When I lived with my parents we would text each other just because we were in different rooms. Modern technology seems to be benefitting our society in many ways, but it is also separating us from our own families.

The quality of communication is lessened and subscribed through digital mediation, rather than facilitated and experienced by the ones talking. For example, a participant wrote:

> When I was having dim sum with my family, I was talking to my mom instead of texting her. The feeling was completely different than if I was to just text her what I said. It was a good feeling, one that I realized I needed more [often].

Because the project required her to forgo texting, she had to verbally communicate with her mother during dinner, rather than texting her. This mother and daughter, despite being at the same table during dinner, generally allow their conversation to be pulled out of physical space and into the digital realm. This choice shows their reluctance to disrupt their digital activity, or detach from digital presence. How often do people feel distracted from others, simply because they want to return to their devices? Participants describe how their friends, children, and other relatives seem to prefer being in digital space, rather than being physically present, even during social events.

> At this time I did the usual, had dinner with my family. That's when I realized that they too check their phone every 5 minutes!! I felt like the parent telling them to put their phone down and eat.
>
> It was nice going out with my parents without having my cell phone on me. I felt a little weird at the same time because my mom was on her phone and I thought it was pretty rude so I told her to get off her phone and she did. It made me realize that it is rude to use my phone when I go out to eat with people.

Eating while using the phone has become common practice for a significant number of people. Pan (2012) reported that a third (32%) of diners use social media while eating instead of talking to family and friends; for Millennials it is almost half (47%).

Quality of Communication: Quality of Relationships

Social interaction can be assessed in terms of quantity—number of likes, how many have friended someone on Facebook, how many followers a person has on Twitter—or it can be assessed in terms of quality. Digital presence has had a significant impact on the quality of shared communication, and the types of relationships being developed. The authors argue that it is imperative to think about the manner in which relationships unfold, and the intensity or "meaningfulness" of the bond. This can be seen in the insights of participants who reflected on the relative strength of social media friends, which might involve a "false sense of connection," as well as the overall value of being more interpersonal during physical presence.

> [A relationship that] lacks a face-to-face interaction … makes it hard for people to determine if it is a "real meaningful relationship" or a "casual relationship …" I still have a meaningful relationship with some of my friends that I haven't seen in years but at one point were really good friends. I feel social media helps one find old friends and [newly] connect with others, but it may not be a real meaningful relationship where they are your main group of friends … we should be able to determine if the relationship is a "meaningful relationship fostered in the real world" verses the "numerous casual relationships [fostered] through social media."
>
> In terms of my relationships this experience has taught me that living in this world influenced by technology we often miss out on a lot. For example, you could talk to a stranger instead of listening to your iPod on your way to work or school, and these are real relationships.
>
> Throughout the day, it is very easy to be distracted by technology. In that case, one could say that young people especially talk less to others because they use only technology communicate … I feel that too much technology can have negative effects. I believe this is serious, because it is through communication where one builds confidence and gains communication skills. If people are on their phones every chance they get, the less time they are actually communicating. When we use social media, email, and text messaging, we think we are connecting with others when we actually are not … I was reminded that an actual conversation there was more of a connection and one uses their emotions more.

In fact, a significant number of the participants (39%) indicated that they became more aware of the importance of in-person interaction and communication while sharing physical space. This awareness was particularly clear in family relationships. This participant expressed it well:

> The best part of my day, however, was sitting in my parents' room and talking to my mother. We talked over an hour about family issues and dramas. I had a lot of fun catching up with her. It taught me how much technology takes my time away from what really matters, family.

Beyond Communication—What to Do When Offline

Aside from face-to-face interactivity, what do Millennials do when not on their devices? Actually, for slightly over a third (36%) of these participants, the answer, for at least part of the twelve hours was "nothing." Surprisingly, when giving up digital presence, many Millennials described being unable to think of what to do with their time. However, the more typical responses during the twelve-hour span involved heightened productivity of some type: almost half (48%) did household chores; over a third studied (39%); over a quarter (28%) engaged in healthy physical activity (like going to the gym). For example:

> One positive outcome of this experience, other than learning how technology dependent I am, was that I was left with no choice but to be productive. I was able to clean and organize my room, which is something that was overdue.
>
> If we cannot play games on our phone…what else is there possibly left to do? CHORES! As horrible as it sounds, I probably never would have cleaned out my cabinets if it was not for this … (my mother thanks you). I was finally forced to get rid of clothes I no longer wear and make room for any future shopping … I think the perception of technology is that it is the only option we have.
>
> Personally, I thought there would be nothing for me to possibly do during this [twelve hours]. I figured I would be cooped up in my room staring at the walls, and I think that's how everyone else sees it.
>
> Throughout these twelve hours I felt that my time was spent more productively than usual … I was able to get to work early without having to speed to make it on time. I was able to realize … I over use modern technology and I should keep it to a minimum … Maybe now I will try not to use technology so frequently like I usually do in my daily routine.

Although this seems like a rosy picture, many of these productive Millennials would lose steam as soon as the task was completed. It seems that rather than complete abandonment of digital presence, the ideal would be a heightened awareness of physical presence and a greater balance between digital space and physical space, particularly in terms of multi-presence. One-third (32%) of the participants mentioned coming to this conclusion themselves.

> I realized that I can find a more efficient and productive way to use my time, rather than waste it. In the end, modern day technology has its advantages and disadvantages, but it is up to us as individuals as to how frequently we interact with it and whether the motives are for productivity, or just as a means to hold off on real and significant tasks.

To Be Alone or Not To Be Alone

In addition to feeling more connected to others while communicating through physical presence, participants have mentioned the importance of being comfortable with solitude. Being with one's own thoughts and doing nothing but reflecting on one's own being is something that can indeed benefit everyone. Of course, it is a challenge to engage in introspection amidst a multitude of distractions, as this journal entry expresses:

> The relationship that had a bigger impact [while offline] is the relationship I have with myself. Honestly, this helped me see things more positively. I am never really in the moment. If I am not listening to music, I am on my computer. I feel like I need to be with myself and my thoughts, but I must give up some of the time I spend with all these electronic devices … This helped me look at my life in a different way. Even though, yes, technology is almost unavoidable now, we must really spend time with ourselves, and others, to develop meaningful relationships.

Escape into Digital Space: Over-Reliance

It is difficult for anyone who is immersed in long-established routines to critically evaluate them. Participants, by experiencing this break from the digital presence that is ingrained in their daily lives were able to distance themselves and reconsider the level of their engagement.

> I learned that I am a bit too dependent on my cell phone and the Internet, in order to get through my day. I realized that most of the stress I carry, I tend to relieve by literally wasting time replying to messages through Facebook, or scrolling through Instagram via this small technological device, which I call my cell phone.
>
> I wrote down in my hourly journals that if I had the choice, I would be on social media. This is where I realized that I can probably finish most of my schoolwork within these two hours which at times I tend to dedicate for meaningless social media browsing. Overall, it is difficult to accept the fact that I have become so dependent on modern day technology. I truly thought I was far from engaging with social media in this manner, but this is my newly found reality. I'm a bit disappointed in myself, but it is a wakeup call for me.

It is not uncommon for people to be unaware of how much they rely on technology. It has become so integrated into the routines of daily life, is so mobile, and can be consumed in any idle moment that people often don't realize how often they use their technological devices.

> As I gave up technology use for twelve hours I felt like it was the hardest thing to do … I felt like there was something missing in my life. I became really anxious because I wasn't able to go on Instagram while sitting here at home … I can now say that I am dependent on modern technology, especially when I have nothing to do and I'm bored.

Perhaps the most detailed picture is drawn out in this journal entry:

> This experience taught me a lot about how I use technology in my everyday life. I wake up to my cell phone's alarm and check my cell phone for calls, texts, emails, and Facebook updates … When I got to class, I realized how much I rely on the digital recorder to take my notes for me. My hand was cramping up from taking notes as fast as I could with a pencil. Then during break, it was so difficult to not call my girlfriend and talk to her. I have called her on all my school and work breaks for the past ten years. It felt like something was missing at that moment … Having no technology to occupy my time, I was able to catch up my class readings. This taught me about how much time technology takes away from me on a daily basis.

A significant number of the participants (40%) reported becoming aware of over-reliance on being digitally present.

What Now?

There it is. It is evident that people spend more time physically communicating, getting things done, and interacting when more fully engaged in their physical presence. Without the digital distractions, relationships are made, strengthened, and in some cases, saved. Interaction is more complete and more mindful while sharing physical presence. Without the presence of their digital devices, people feel more confident and empowered to interact with others. Moreover, some discover that they gain insight from being alone with their thoughts. Digital presence can be a distraction from interactions taking place in shared physical space. At times it significantly replaces in-person interaction. It can suppress productivity and undermine family relationships. In fact, there was a significant upswing in productive work among participants: almost half (48%) did more around the house, while over a third (39%) increased their study time. Many (29%) mentioned a new awareness of the need to better manage their time with digital presence in general. More than a quarter (27%) specifically mentioned improved time with family, while some (14%) gave examples of positive family reactions when participants refrain from excessive co-presence, or when the family joined in to eliminate multi-presence. Overall, the lesson learned was the need to consciously decide on the relative balance among full physical presence and co-presence, and to carefully consider the context for multi-presence.

References

Blatchford, Emily. 2017. "How Often Should You Text Your Partner?" *HuffPost*, Aug 9. Retrieved February 17, 2018 (https://www.huffingtonpost.com.au/2017/09/07/how-often-should-you-text-your-partner_a_23198283/).

Center for the Digital Future. 2012. "Special Report: America at the Digital Turning Point." USC Annenberg, January. Retrieved April 20, 2017 (http://www.digitalcenter.org/reports/).

Gasser, Urs and John Palfrey. 2009. "Mastering Multitasking." *Educational Leadership*, 66(6):14–19.

Hall, Jeffrey A. and Nancy K. Baym. 2011. "Calling and Texting (Too Much): Mobile Maintenance Expectations, (Over)dependence, Entrapment, and Friendship Satisfaction." *New Media and Society* 14(2):316–331.

Pan, Joann. 2012. "Tweets at the Tables? More of Us Mix Social Media and Food." *Mashable.com*, February 29. Retrieved April 20, 2018 (https://mashable.com/2012/02/29/social-media-andfood/#wBleiTsRO5qR).

University of North Carolina News. 2016. "Overusing Social Media as an Adolescent May Hurt Your Love Life Later On," June 21. Retrieved April 20, 2017 (https://uncnews.unc.edu/2016/06/21/overusing-social-media-adolescent-may-hurt-love-life-later/).

9

INTERFACING CONFLICT

Advice Columns and Digital Life

Cristina Bodinger-deUriarte, Gunnar Valgeirsson and Sasha Santhoff

The Origin of Advice Columns

Advice columns have a long history and can be traced all the way back to the seventeenth century. In 1691, English book publisher John Dunton had an extra-marital relationship that he deeply regretted (Kent 1979). This transgression consumed his thoughts and he desperately wanted guidance on the proper way to deal with his problematic situation; he longed even more for anonymity. At the time, infidelity was a serious social taboo, and Dunton, who usually thought of himself as virtuous, was worried about the scandal that would result if his adulterous behavior came to light. Wracked with anxiety and guilt, he considered confessing his sins "to some divine," to relieve his conscience (Nichols [1818] 2012). Instead, he shunned the church, but approached the public.

Dunton realized that he could not be the only one in a dire position who desired secret council. Returning home, he invented what has come to be known as the "advice column." Dunton published the *Athenian Mercury*, the first paper to encourage audience participation. The title was inspired by a passage in the New Testament: "The one amusement the Athenians and the foreigners living there seem to have is to discuss and listen to the latest ideas." *Athenian Mercury* readers were promised the opportunity to anonymously send in questions where they would be answered by a panel of twelve learned men—better known as the Athenian Society.

The Athenian Society, however, did not exist. It was nothing more than a marketing ploy. Instead, Dunton and a few other men answered all questions. When the number of questions became overwhelming, they added two more men. These six remained the advising body, however guest advisors sometimes contributed.

The *Athenian Mercury* became wildly successful, especially with women, who often asked about courtship and marriage. Consequently, Dunton employed women writers to help out with "feminine" topics. Within two years, the *Ladies' Mercury* was published. It was only in print for four weeks, and consisted of a single sheet of paper devoted solely to women's questions. It is considered by many to be the first periodical written solely for women. This earned Dunton the label of "feminist" by many academics (Kent 1979). An ironic honor considering that this man was inspired to create the advice column after cheating on his wife, to advise men in similar circumstances. Further, it evolved to the point where he was able to profit from the problems of his female readers.

The Persistence of Advice Columns

Why do people ask strangers for advice? According to NPR pop culture maven, Linda Holmes, they may be looking for validation while trying to seem objective.

> We've convinced ourselves that we really do want to know what an unbiased third party would think … In that sense, people like advice columns precisely *because* the person is a stranger. Dear Abby had no loyalty to you, or your boyfriend, or your best friend, or your boss. Dear Abby, ideally, told it like it was (Holmes 2013).

However, it is important that advice-seekers believe columnists duly consider the problems they write about. For the act of advice-seeking to be legitimate, the advice-giver cannot seem to automatically support everyone who writes in.

> Asking for advice doesn't make you the hero. In fact, a good advice columnist knows when to say, "You know what? It really sounds like the problem is you." … A willingness to call out the reader, and not to coddle everyone who writes in, can lend a columnist credibility and seriousness. (Holmes 2013)

Cultural trends can have a short shelf life, but advice columns have lasted for centuries and are proliferating. There are still traditional newspaper and radio venues with printed columns and call-in programs. However, potential advisors no longer have to compete for only one or two advice columnist positions contracted to publish in hardcopy newspapers or magazines. There are online versions of the newspapers, advice websites, blogs, and podcasts. Most well-known advice-providers show up on multiple venues. Digital life not only provokes letters from advice-seekers, it also provides unlimited advisement forums. This has led to a plethora of customized advisement venues in digital space. *Ask a Manager*'s Alice Green advises on workplace conflict. *Ms. Vicki* writes a column that specifically caters to the significant others of military personnel. *Ask Bear* can advise a broad audience, but specializes in LGBTQ advice-seekers. Similarly, *Captain Awkward* can address the masses, but has specialty features for "geeks." Among the best known of the specialists, *Miss Manners* helps with all things etiquette (and netiquette).

Advice column issues continue to span all forms of relationships—those with intimates, immediate family and more distant kin, friends, co-workers, employers, and neighbors. To a lesser degree, they include general etiquette in fleeting relationships with service personnel or random others. In this sense nothing has changed since the first Dear Abby. However, the *venue* for conflict, and the activities that provoke conflict, have significantly changed in the digital age. The areas of conflict used to primarily involve interactions in one's home, during visits to one another's homes, at social events, during dates, and in the workplace. Arenas of conflict now include posts and comments on one's Facebook and Snapchat, as well as other digital interactivity (or lack thereof), digital mediation of social events, online dating etiquette, netiquette during offline dating, possible workplace consequences of online actions, telecommuting conflicts, and so on.

Considering how thoroughly Americans integrate their digital presence into their daily routines and practices, if advice columns reflect the interpersonal concerns of their readers, online issues should be well-represented, particularly on Facebook, the most popular social media platform. This has, in fact, transpired. Even five years ago, Amy Dickinson (2013) reported that one-third of the letters she received about online conflict involved Facebook. One might predict that advice-seeking for the digitally distressed would cluster around areas with high post rates. The most recent study of social media posts includes four categories with interactive components relevant to advice column content. The proportion of social media users who posted in these four popular advisement areas follows (see Table 9.1):

Table 9.1 How Many Social Media Users Post on These Topics

Work	*Children*	*Friends*	*Family (not kids)*
7%	18%	26%	34%

Source: Data from Statista 2018c.

The letters that address digital life are similarly concerned with family, particularly spouses, parents and in-laws. Friends are also frequently featured. When letters feature offspring, they tend to be adults. Grandparents and stepparents are more likely to mention young folks than are their parents. Relatively few letters focus on work-related issues aside from office romances. The advice column letters clearly align with the topics most often posted.

Digital Distress

This chapter argues that advice columns provide a microcosm of social norming at the individual level. They present a record of historical cohort shifts in interpersonal concerns, and a record of changing ideas about recurring interpersonal issues. They also provide a touchstone for contemporaries who seek to understand current expectations. Finally, they may, to some small degree, contribute to the ongoing change of social norms by increasing the visibility of conflicted issues and modeling empathetic responses and enlightened outlooks.

Digital-age advice-column letters do, indeed, reflect netiquette issues, as well as deeper concerns among netizens. The letters, however, are not focused solely on digitally-mediated experiences. Letters frequently address problems that interconnect the writer's physical and digital presence environments. Letters focus on widely ranging issues with headings such as "Husband's iPhone addiction Leaves Wife Lonely," and "Inattentive Facebooker Misses News of Friend's Tragedy, Commits Faux Pas," and "Protecting Kids' Privacy Online."

Advice column letters increasingly have digital themes. For this study, all recent advice column letters with digital themes in three of the most widely read mainstream advice columns were categorized. This yielded 255 letters with direct references to the role of digital life in the writer's current difficulty. However, 59 of these, just under a quarter (23%), focused on problems in intimate relationships that now have a digital component, but are not significantly different from pre-digital intimate relationship issues. This most frequently this involves accusations of infidelity from a partner's suspicious texts. This is not significantly different than accusations of infidelity from suspicious notes or credit card charges. The other 196 are more tied to the specific nature of digital life, or dynamics that are escalated or play out differently due to digital abilities (see Table 9.2):

Table 9.2 Conflict Categories

196 letters	*Letters per Category*	
Boundary Violations	45	23%
Online Dating Issues	38	20%
Friending/Unfriending/Ghosting Issues	32	16%
Cyberbullying	26	13%
Outcome Anxiety	24	12%
Netiquette	16	8%
Miscellaneous	15	8%
TOTAL	196	100%

It must be noted that these proportions do not necessarily reflect the frequency with which these issues arise among letter writers. Editors and columnists are the gatekeepers and choose a small subset from numerous letters received. They may choose those they feel are most important, interesting, or entertaining for their audience. There is no way of ascertaining the representativeness of the letters making it into public view. For example, in their heyday, weekly letters to Ann Landers column had reached 14,000 and Dear Abby received even more. Today, even less prominent, but well-known columns like "Dear Prudence" receive an average of 400 letters a week. Only a small number of them receive responses across her three formats: a weekly column, a live chat twice a week, and recorded video responses (Petronzio 2015). *Ask a Manager* receives 350 workplace questions weekly and receives about 2.5 million visits a year (Green 2017). That said, the fact that these *are* selected indicates the degree to which editors and columnists feel they will relate to the audience—perhaps an indication of their relative importance for the broader public.

Intimate Relationships

Digitally-based disruptions in intimate relationships were the most common digital problems shared by columnists. This may either indicate that disproportionately high number of letters addressed this, or it might be that intimate relationship conflicts have always been advisor or editor favorites. It may simply reflect that the category is broad; it subsumes a number of subcategories that could have been separated out. Among other issues, letters commonly include the following: loss of an intimate's attention due to online excess; over-texting or cyberstalking an intimate; suspicious social media checking, discomfort over an intimate receiving likes or comments from a third party on photos or other posts (or giving them to a third party); discovering that someone is sending suggestive texts to one's intimate; digital indications of actual or impending infidelity. Selected excerpts illustrate each of the above-mentioned issues.

Letters involving an intimate who is increasingly unapproachable due to competing digital activity are sent from both men and women. Advice-seekers try to ascertain to what degree they are expected to accommodate an intimate's digital activity when they (or their children) are physically present.

> A large portion of conversation in our home has been replaced with the time my wife spends on her cellphone … games and Facebook that absorb hours of her time. We used to spend evenings working on special projects together, but that is the exception now and no longer the rule (Van Buren 2017a).

As of four years ago, studies already showed that 18 percent of millennial couples have argued with a partner about the amount of time one of them spent online, more than twice the 8 percent rate of couples across all age groups (Lenhart and Duggan 2014).

In a related sub-category, but from the opposite end, letters express frustration with an intimate who is demanding their attention too much by over-texting them, particularly when it interferes with work.

> I am on the phone for a living. I cannot stand being on it more than I must be. He calls and/or texts me up to 12 times a day. I can't stand it. Even when I'm busy or give him a time certain when I will call him back, he beeps in before I have the chance (Van Buren 2017f).

Studies have confirmed this as a serious source of marital frustration. Escalating expectations for frequent and immediate texting replies leads to feelings of entrapment and relationship dissatisfaction (Hall and Baym 2011). Experts found that establishing mutually agreed upon rules about frequency and duration of texting exchanges increased relationship satisfaction (Miller-Ott, Kelly, and Duran 2012).

Frustration is also expressed when an intimate becomes engaged in digitally documenting perceived relationship threats or suspected transgressions. This sometimes includes demanding passwords and access to all social media accounts.

> [My husband] has gone so far as to create secret social media accounts to monitor my daily activity. He also monitors [my friend's]. It became somewhat of an obsession. He has been checking several times a day and screen captures anything he deems inappropriate with ANY of my friends (Dickinson 2016a).

This type of chronic jealousy and suspicious checking behavior has always been a relationship destroyer. Digital platforms enable this form of jealousy, and make a wider range of checking and tracking available than previously possible. This type of digitally-enabled harassment, particularly when the target feels threatened, is considered cyberstalking.

Social media also provides more venues for provoking jealousy. A study by Pew Research Center found that "8% [of Millennials] say they have been upset by something their partner was doing online;" for older couples, it was half that percentage (Lenhart and Duggan 2014).

> [My boyfriend] keeps contacting women that he dated before we got together. He comments on their Facebook pictures, telling them how gorgeous they are ... he says he loves me but keeps

> adding women he doesn't even know personally to his Facebook list … Every time I say something he gets defensive and starts a fight saying I'm jealous. I know I'm jealous (Dickinson 2016b).

Perhaps more frustrating is the inability to control who interacts with a significant other on social media sites.

> Guys from the gym [my wife works at] have added her on Facebook and send her messages. They like all of her posts and pictures. I work out there … I see her laughing and joking with them … she just says they are my insecurity issues, and I need to deal with them … I check her Facebook page constantly to see if she has added any new guys and see what comments they are leaving. I know it's not healthy (Van Buren 2016b)

One recurring issue is those upset by texts received by an intimate from a third party. The problematic behavior is frequently something everyone would admit was inappropriate face-to-face.

> [My wife] loves her job … She isn't someone who makes friends easily so when she does she keeps hold of them. She has recently become good friends with one of her male work colleagues who … sends her texts every day … I have seen some of the texts … Asking her …if she can wear a particular dress that he likes … asking if she has sex toys … she shrugs it off saying … there's nothing in it. I am right by thinking that another guy shouldn't be texting anything like that to another man's wife let alone a work colleague? Right? (Mumsnet 2014).

Recipients may be able to dismiss such texts because there is no awkward face-to-face component and they can be ignored relatively easily, especially by those adverse to conflict. Frequently, however, the result is conflict with the intimate over failure to put a stop to unwarranted texts.

This leaves the area of infidelity. Infidelity is what inspired the unfaithful Dunton to create the first advice column. It has remained a staple of advice columns for over 320 years. Abigail Van Buren once referred to it as "the one activity that provides me with more column material than any other" (Van Buren 1981). Consider that Facebook came out eight years after the following letter:

> The World Wide Web and the Internet … can be home wreckers. My wife of 18 years left me for a guy she met on-line … There was no hint of trouble until she began to stay up until 3 a.m. chatting on-line with a man … Last week, when I came home

> from work, I found a note saying she went to meet [him] … She promised to call me "soon" (Landers 1996).

Not all forms of digitally facilitated "cheating" involve "running off together," or even direct contact. With the advent of social media and messaging technologies, one can be in a secret, deeply engaged, and committed relationship in cyberspace that undermines another established relationship in the physical space that they directly inhabit with an intimate. This is the relatively new phenomenon of an "emotional affair" or "online infidelity." Marriage and family therapists have reported more couples experiencing problems due to actual or suspected or online infidelity. A licensed psychologist in Maine reported that two-thirds of the couples seeking his help do so for issues related to online infidelity (Melby 2010).

Boundary Violations

Traditionally, when people raised in the same culture are face-to-face, they have common understandings about what is appropriate to say to one another, what types of gossip about others is considered "going too far," what types of information can be passed along, and whether a rant or a religious or political diatribe is appropriate. The same is true of photos of other people—consider the old example of a parent showing baby photos to a teenager's new "date." These unwritten social norms are not generally problematic; boundaries are understood and, for the most part, respected. This may be because the cost of ignoring such norms is often social awkwardness and embarrassment. Sharing physical space with one another in such circumstances can be quite uncomfortable for all involved.

Understandings about shared information and images may change as the way we share information changes, however. Social media expectations include the frequent sharing of images and snippets of information. The worthiness of these posts is assessed through quick, clickable responses like "Likes," or actual comments. Frequent posting creates pressure to continue providing posts. Posting to amuse or engage others may diminish ones sensitivity to who may be bothered by the timing or content of the post. Mutually understood and well-established boundaries, may no longer seem so. The social cues available in shared physical space to alert us to a gaffe are not as visceral in digital space exchanges. Boundary violations arise in disagreements about sharing information as one set of expectations conflicts with another (Ward 2006).

Letters categorized as boundary violations include issues with online friends and relatives posting images of the writer or the writer's children without permission, posts attempting to garner third-party support in a dispute, posting family "secrets" of living relatives, and posting information ahead of those who should be the ones to first share. Now that people routinely illustrate their daily lives, this can blur the lines of what images are appropriate to capture and post, leading some to believe they are entitled to anything in view.

> I asked a close female friend to help me prepare a gift for my husband for our fifth wedding anniversary … a photo album of sexy nude photos of myself … my husband loved it … I recently found out on social media that my girlfriend's husband viewed and copied my private photos, and shared them online with mutual male friends of ours. I'm devastated (Van Buren 2017e).
>
> We attended a family function and a guest at the party took pictures of me and my children. He proudly showed me the images and told me that he had posted them on his Facebook account … I kindly told him to please remove the images off his account. I was shocked by his reaction; he became upset and made a huge scene at the party, but I did not back down (Dickenson 2017).
>
> My mother has been using Facebook … I am not her "friend" on Facebook, but if I look up her name I can see dozens of photos of me all over her page. I really don't want her plastering my picture everywhere … her privacy settings are weak and anyone has access to the photos … My personal page is very private, and I only have friends that I actually know on my page … I have asked her to take down many of the photos but she refuses … because she is my parent does she "own" images of me? … I am legally an adult. What should I do? (Dickinson 2014b)

The situations get complicated when family get involved and repost the posts. Once on a media site aren't they free game? Netiquette says it may depend on what social media site privacy choices were selected by the person who originally posted the photos. A grandparent wrote to ask why her grandkids were upset when she posted to her Facebook the screen shots of their posted photos. This is a boundary violation because

> "They've selectively given permission to the people they want to see the photos. What you're doing is violating that trust they've given you" (Vernon 2014).

Online Dating Issues

These letters include negotiating the transition from online to first offline meeting as well as how to cope with disappointing results (being rejected or withdrawing after meeting face-to-face). They sometimes express anxiety over having misrepresented themselves or they worry that an anticipated partner is misrepresenting something. Fear of misrepresentation underlies a significant number of letters expressing the safety concerns of parents and friends. Advice-seekers also ask how to push back against parental limits and suspicions.

> I think online dating is a great idea in theory. But I have anxiety … What if my anxiety makes me shaky? What if I can't talk at all? What if there's no chemistry? … The logical part of my brain knows none of these things matters … and if it doesn't work out we'll both just move on with our lives … I've been chatting with someone online who I look forward to meeting, but anxiety persists (Hax 2018b).
>
> I made a joke Tinder account with friends, with a fake age, job, etc … . Turns out, I really hit it off with a guy … The thing is, he's 23 and I'm 17 (Dickinson 2018b).
>
> I met this guy through a dating site … We talk almost every day, and have Skyped a few times … I'm not sure if he's telling me the truth. We live in different countries, and for all I know, he could have a girlfriend … I also don't know how to tell my parents. They don't support online dating (Van Buren 2017b).

Online dating issues are sure to arise when so many people meet this way. Online dating is a billion dollar industry. There are venues primarily introducing those who just want to "hook-up," those that tout their marriage rates, and everything in between. After 2005 one-third of marriages in the US were between partners who met online; of these couples, nearly half met through online dating sites (Cacioppo et al. 2013).

There are relatively few advice-seeking letters from established couples indicating that the writers initially met online. This may mean that most of the letter writers met in physical space, which would reflect findings from studies showing that the marriages of those who "met their spouse on-line were rated as more satisfying than marriages that began in an off-line meeting." Further, it could also be affected by the fact that "marriages that began in an on-line meeting were less likely to end in separation or divorce than marriages that began in an off-line venue" (Cacioppo et al. 2013). Then again, it may simply be that writers do not provide the information. Consequently, it is not possible to identify what proportion of advice-seekers are in relationships that originated online, compared to those whose relationships originated offline.

Friending, Unfriending and Ghosting Issues

These letters include a wide variety of friending, unfriending, and blocking issues. Who to friend, and how to navigate unfriending, is a growing concern for large numbers of social media users. This extends to employers, co-workers, and clients, as well as relatives, exes, and, pals. Workplace-oriented websites frequently feature articles such as "Consider This Before Friending a Client," and "Six Reasons You Shouldn't Friend Your Employees on Social Media." Friending and unfriending has also taken on legal ramifications, inspiring attorney-written advisory documents such as: "Facebook Friending Your Employees is a Great Way to Get Sued," and "Friending Personnel Problems? Legal Implications of Social Media in the Workplace" (Sanchez and Smith 2011).

The impact of unfriending is the topic of serious study (Sibona 2014) and is also of increasing concern in mental health blogs, columns, and websites. On GoodTherapy.org, the Licensed Marriage and Family Therapist (LMFT) discusses whether friending your ex on Facebook is a good idea (Ruble 2013). Meanwhile, on MentalHelp.net a Licensed Clinical Social Worker (LCSW) addresses unfriending, and warns that:

> Being shunned on social networks is a very real phenomenon. Being ostracized this way results in lowered self-esteem and feelings of depression and of not belonging. Anyone who has participated in an online community support web site can testify to how much emotional power can be stimulated by online relationships (Schwartz 2013).

This corroborates research by Sibona who said that his "study makes clear that unfriending is meaningful and has important psychological consequences for those to whom it occurs" (AAAS 2013). This, of course, extends to ghosting—the sudden and total cessation of all contact and the act of ignoring or blocking all attempts at contact without explanation.

> Ghosting is the ultimate use of the silent treatment, a tactic that has often been viewed by mental health professionals as a form of emotional cruelty. It essentially renders you powerless and leaves you with no opportunity to ask questions or be provided with information that would help you emotionally process the experience. It silences you and prevents you from expressing your emotions and being heard, which is important for maintaining your self-esteem (Vilhauer 2015).

This experience may feel intensified by the number of different contact mediums being ignored—phone, text, and multiple social media sites. As one victim describes it:

> The disregard is insulting. The lack of closure is maddening. You move on, but not before your self-esteem takes a hit. The only thing worse than being broken up with is realizing that someone didn't even consider you worth breaking up with (Vilhauer 2015).

The prevalence of ghosting makes it an expected topic in advice columns since, “in today’s dating culture being ghosted is a phenomenon that approximately 50 percent of all men and women have experienced” (Vilhauer 2015). Given its apparent importance in different social institutions, as well as recent acknowledgement of its potential psychological impact, friending and unfriending should also be represented in advice columns. In fact, friending, unfriending and ghosting accounted for the third most prevalent topic among advice letters concerning digital aspects of relationships.

These letters, however, tend not to represent the full list that Sibona identified as unfriending motives (Schwartz 2013):

1 Polarizing posts usually about politics or religion
2 Inappropriate posts involving sexist, racist remarks
3 Frequent, unimportant posts
4 Boring everyday life posts about children, food, spouses etc.

Advice-seekers most often mention the first two as motivations for unfriending and blocking texts, even among close family members. The Dear Abby response below provides a window on these dynamics.

> Blocking your son from your phones was a mistake. It would have been better to have just told him you would prefer not to discuss politics via text messages… Unblock your phone and let your son know that he hurt your feelings, which is why you did it (Van Buren 2017d).

The letters also reveal that the category of “polarizing posts” has expanded to include a focus on family and friendship-circle dynamics, as well as strife with an intimate:

> I recently got into a spat with my sister about Facebook. We’re both in our mid-20s. She is furious that I don’t add her Facebook friends to my own friend roster on FB and that I deleted one of her friends over a year ago … . It greatly upsets her that I won’t add her friends, even though I don’t know them well and don’t want them to see what’s going on in my life (Dickenson 2015).
>
> There have been covert telephone calls and late-night text messages to a boyfriend from long ago. He’s divorced and found her on Facebook, telling her he needed ‘advice on dating.’… She says she knows she was wrong to let it go on as long as it did, and further communication will stop. Should it include unfriending him on Facebook? (Van Buren 2018b)

Further, unfriending can be a response to perceived sexual harassment or to simply avoid even an offline relationship.

> After she contacted me on a social website, we met and have continued seeing each other for lunch, movies, sporting events … I misinterpreted her attempts to become friends with me. I liked her very much and became attracted to her … a missed communication, an irrational act by me as well as an insensitive comment I made, upset her and now she no longer wants to talk with me … un-friended (Van Buren 2018a).

These letters also include reluctance to add relatives who want to be added on social media, choosing to "unfriend" others due to ideological retaliation, parents expressing concern over their child's online friends, and parents and others who want to be online friends with their child. This is confounded by the absence of shared norms. As *Miss Manners* puts it, "The etiquette around 'unfriending' is still evolving, a process impeded both by the nomenclature and the indirectness of the interaction" (Martin, Martin, and Martin 2018b). The longer it takes to establish common cultural friending and unfriending expectations, the longer advice columns will include such content.

Cyberbullying

Cyberbullying is widely recognized as a social problem. It has reached such a level of importance that an official US government website has a whole section devoted to it.

> Cyberbullying is bullying that takes place over digital devices like cell phones, computers, and tablets. Cyberbullying can occur through SMS, Text, and apps, or online in social media, forums, or gaming where people can view, participate in, or share content. Cyberbullying includes sending, posting, or sharing negative, harmful, false, or mean content about someone else. It can include sharing personal or private information about someone else causing embarrassment or humiliation (US Department of Health and Human Services 2018).

Advice column letters include issues regarding others posting embarrassing images, private information, false information, and antagonistic or cruel remarks.

Unfortunately, the cyberbully is often someone close to the victim.

> We had a fight two weeks ago, and the next thing I knew, I was reading about it on Facebook, all from his point of view … I feel

> hurt after reading what he's posting and ashamed for not being a "good enough" spouse. Am I overly sensitive, or should I confront him about this? (Van Buren 2017c)
>
> My adult niece, 40, posted a very nasty comment about me to a social-media site because of my stance that women and girls should NOT be judged on their looks. This stance was never communicated to her specifically … How do I get over the hurt, since I have no idea where the hate originated? (Hax 2017)

Of course, mainstream advice columns only provide imperfect indicators of social issues, as they do not equally reflect the population. If advice columns published more youth-authored letters, this category for digital distress might be larger. Middle school and high school students are frequent targets. According to the Cyberbullying Research Center, 11 percent in this age group admit to cyberbullying others, while one-third (34%) report having been victims, almost half of them within the month (Statista 2018a). Some are embarrassed to admit being victimized, however. Consider that most teens (88%) and over two-thirds of adults (69%) report witnessing cyberbullying on social media sites (Trend Micro 2018).

Digital victimization leads to real-world pain and, in some cases self-harm. One distinction between offline bullying and cyberbullying is the relatively relentless and inescapable nature of digital harassment, as well as the degree of public exposure suffered by the target. Another differentiating factor is the permanency of cyberbullying content compared to physical space bullying that ends with the end of the bullying encounters. One grandmother sought advice from *Dear Abby*, when her granddaughter claimed that being victimized by cyberbullying left her so depressed she had started to cut herself.

The newest manifestation of cyberbullying is victimizing one's self. "Digital self-harm" is when the targets of the abuse are actually the ones posting the negative online content about themselves (Patchin and Hinduja 2017). Coping with self-hatred is one motivator. Other related factors include sexual orientation, experiences of cyberbullying from others, bullying at school, and depressive symptoms (Ibid). The Cyberbullying Research Center found that 6 percent of the 5,593 teens interviewed had engaged in some form of digital self-harm (Fraga 2018). Males are slightly more likely cyberbully themselves than are females (7% versus 5%) according to Patchin and Hinduja (2017). Digital self-harm is, perhaps too new a phenomenon to have arisen in the advice columns. However, if advice columns are truly reflective of the concerns of their time, concerns around digital self-harm can be expected among advice letters in the near future.

Outcome Anxiety

These letters are concerned with the possible outcomes of someone else's social media posts—for the one who posted (or their loved ones). The anxiety here is two-fold. On one hand, they are concerned about the consequences of the post. On the other hand, they are worried about whether they should communicate these concerns, and if so, how to go about it.

Are such concerns overly dramatic? Not according to experts at Bernstein Crisis Management and Trend Labs Security Intelligence. Oversharing backlash commonly takes six forms: (1) criminal "social engineering" schemes—cyberfraud and offline scams—that build on social media posts; (2) cybercrime identity theft from personal profiles and posts—often by resetting user passwords using the information gathered through SNSs; (3) cyberbullying that uses personal information against posters; (4) targeted advertising—especially when one in four users "like" targeted ads; (5) offline crimes like burglary and stalking when users post vacation plans and whereabouts—4.8 million people have revealed their vacation plans online; and (6) the one advice-seekers are most concerned about on behalf of their friends—damaged reputations, given that three-quarters (75%) of hiring managers check social media (Trend Micro 2018).

Typically, anxieties are triggered when friends or relations overshare or post potentially risky or self-sabotaging comments or photos on social media sites. At what point does oversharing have consequences? Do you warn an adult of the potential repercussions of posting provocative selfies? Drunk posting? Offensive rants? These excerpts illustrate this type of advice-seeking letter.

> We are more concerned that her [fake] online persona doesn't end with just the posting of photos, and that she may actually be pretending to be someone else and may be developing online relationships based on that. This could seriously harm (or ruin) her current relationship or her job. Should we confront her, and, if so, how do we go about it without ruining our relationship with her? (Dickinson 2018a)
>
> A friend and former co-worker of mine recently changed her Facebook profile picture to a full body shot of herself lying in bed, wearing a camisole and panties and staring seductively up into the camera … [She's] doing well in school for a future career in a highly respected medical field … these photos will almost certainly come back to haunt her … Is there a tactful way I can address my concerns with her? (Dickinson 2014a)

Of course, if most social media users followed the advice of Miss Manners, the motivation for oversharing might fade away:

> It is now the unpublicized life that is thought to be not worth living. Miss Manners would be wary of continuing to watch the posts of people who believe that the world wants to share their most intimate moments (Martin, Martin, and Martin 2018a).

Netiquette

One of the difficulties in overall netiquette relates to the fact that there are no clear and well-established unwritten rules. This kind of ambiguity leaves the expectations unexplained and the ideas about "problematic" too individualized to provide any guidance (Hertlein and Stevenson 2010). Many of the issues involve appropriate communication: When to use e-cards versus paper cards; if emailed invitations get snail mail thank you notes; if it is proper to be invited to weddings over email; whether one has the right to be upset that a sister used email to break the news of their mother's death.

Other issues involve balancing between the expectations of those online and those sharing the writer's physical space. This includes who to greet first on entering a room, the spouse and the person she is Skyping with, or the son who has run over to welcome daddy home? When can one take over the family television to pursue gaming? Can one check their phone at the theater during the very brief set change (not intermission)? How to handle being ignored in person while the online friends get all the attention.

> I had my friend and her husband over for dinner with my boyfriend and me. For most of the night she was on the phone Facebooking and texting pictures of my dinner table to people I don't bother with. There was absolutely no conversation between the two of us that night. When I said something about her being on the phone, her answer was that she was answering her Facebook messages (Van Buren 2016a).

Some of these issues can become quite socially fraught. What is the etiquette around how frequently one should check a Facebook friend's posts?

> We were friends, but I had muted her [Facebook] postings since she posted quite frequently … she suffered a late (seven months in) miscarriage. She had many detailed posts of her anguish … My question is, is it worth bringing up months after the fact? … And what is my responsibility to be aware of people's personal issues if we are social media friends but I do not actually pay attention to their posts? (Hax 2018a)

These more general matters around conflicting definitions of politeness or civil expectations are relatively minimal compared to the other issues people are grappling with as they come to terms with their interwoven digital and physical presence.

Conclusion

Advice columns continue to thrive despite technological changes that upended their original media delivery systems. The current study indicates that advice columns remain a touchstone of current cultural practice. The issues that motivate people to write to advice columnists continue to evolve as society and technology evolve–so do the solutions offered, and the avenues through which advice-seekers and advisors connect. As digitally-mediated relationships, communications, and connectivity become routine, digitally related conflicts and problems regularly surface.

No one would argue that social media has little impact on people's lives. For social and behavioral scientists, however, the points of impact, their breadth, depth, and saliency are most important. Advice columns paint a picture of the range of interpersonal issues in broad strokes. Also, few other sources capture the depth of sentiments and strength of reactions people experience when navigating the pathways of digital life. Columns also capture confusion and conflict over social expectations attached to newer, digitally-mediated behaviors.

Of course, mainstream advice columns only provide imperfect indicators of social issues. Advice columns reflect social norms, in part, by making norm violations prominent. Further, advice columns are predicated on dysfunction. Consequently, the prevalence of these letters in no way indicates the *degree* of its prevalence in society.

Advice columns also do not equally reflect the population. These advice column letters primarily represent adults, with occasional inclusion of emerging adults. If advice columns published more youth-authored letters, the distribution across categories might be quite different. For example, cyberbullying might be more represented and gaming and sexting would probably break out of the "miscellaneous other" category. The predominance of Facebook would be less pronounced as well, as other digital venues would share in the spotlight. Those aged 18–24 are also highly involved with YouTube (71%), Instagram (58%), Twitter (47%), and Snapchat (45%)—venues in which conflict also transpires—whereas the ages 25–49 group averages a 24 percent lower digital presence on these sites (Statista 2018b). Although not elaborated enough or systematic enough to provide robust data on particular digital-life topics, advice columns provide an informative orientation and foundational indicators for the direction of such research.

The question that remains is whether advice columns might also be a venue for social change. Certainly, they can reinforce particular values in their responses. They can also be used to represent a position in a conflicted social landscape. For example, in 2013, when Amy Dickenson wrote a chastising response to a mother who wanted her son to just choose to stop being gay, it went viral. Dickenson's response was framed in such a way that it was seized as a means of promoting a valued viewpoint: it was posted on *Upworthy* and *Buzzfeed* and tweeted by the actor and social activist George Takei (who has 5 million followers) and led to an interview on GoPride.com, an LGBTQ website. Advice columns today have expanded their reach beyond anything possible in the print era. With millions of potential views, and expanding audiences through re-uploading, and tweeting, and responders to trending notifications, digital advice columns have the power to aim a spotlight at particular issues, by what letters they feature, and to engage in consciousness-raising by the way they frame responses.

References

American Association for the Advancement of Science (AAAS). 2013. "Study Shows Facebook Unfriending has Real Life Consequences." *Eureka Alert*, February 3. Retrieved May 7, 2018 (https://www.eurekalert.org/pub_releases/2013-02/uocd-ssf020413.php).

Cacioppo, John T., Stephanie Cacioppo, Gian C. Gonzaga, Elizabeth L. Ogburn, and Tyler J. VanderWeele. 2013. "Marital Satisfaction and Break-Ups Differ Across On-Line and Off-Line Venues." *Proceedings of the National Academy of Sciences (PNAS)* 110(25):10135–10140. June 18. Retrieved May 17, 2018 (http://www.pnas.org/content/110/25/10135).

Dickinson, Amy. 2018a. "Family Worries that Niece is Online 'Catfish'." *Ask Amy: Chicago Tribune*, May 8. Retrieved May 8, 2018 (http://www.chicagotribune.com/lifestyles/askamy/ct-ask-amy-ae-0802-20170802-story.html).

Dickinson, Amy. 2018b. "This Guy on Tinder Says It's OK that I'm 17: Should I Go For It?" *Ask Amy: The Mercury News*, May 11. Retrieved May 10, 2018 (https://www.mercurynews.com/2018/05/11/wife-wonders-if-she-should-leave-boring-marriage/).

Dickinson, Amy. 2017. "Parent Refuses to Let Others Post Kids' Photos." *Ask Amy: The Denver Post*, December 7. Retrieved May 7, 2018 (https://www.denverpost.com/2017/12/07/ask-amy-parent-refuses-to-let-others-post-kids-photos/).

Dickinson, Amy. 2016a. "Boyfriend's Social Surveillance Spells Trouble." *The Chicago Tribune*, August 28. Retrieved May 15, 2018 (http://www.chicagotribune.com/lifestyles/askamy/ct-ask-amy-sun-0828-20160828-column.html).

Dickinson, Amy. 2016b. "Sorority Rush a Bust." *The Chicago Tribune*, October 11. Retrieved May 15, 2018 (http://www.chicagotribune.com/lifestyles/askamy/ct-ask-amy-ae-1011-20161011-column.html).

Dickinson, Amy. 2015. "Ask Amy: Sisters Engage in a 'Friending' War." *Ask Amy: Washington Post*, December 10. Retrieved May 7, 2018 (https://www.washingtonpost.com/lifestyle/style/ask-amy-sisters-engage-in-a-friending-war/2015/12/02/bddf64dc-97a2-11e5-94f0-9eeaff906ef3_story.html?utm_term=.a1843318d7fc).

Dickinson, Amy. 2014a. "Friend Worries About Racy Facebook Profile Pic." *Ask Amy: Chicago Tribune*, March 4. Retrieved May 10, 2018 (http://articles.chicagotribune.com/2014-03-04/features/ct-racy-facebook-photo-20140304_1_friend-worries-racy-facebook).

Dickinson, Amy. 2014b. "Mom Violates Her Child's Facebook Privacy." *Chicago Tribune*, July 3. Retrieved May 15, 2018 (http://articles.chicagotribune.com/2014-07-03/features/ct-mom-violates-facebook-privacy-20140703_1_facebook-privacy-privacy-settings-personal-page).

Dickinson, Amy. 2013. *Ask Amy: Advice for Better Living*. Evanston, IL: Agate Midway.

Fraga, Juli. 2018. "When Teens Cyberbully Themselves." *Mindshift KQED News*, April 22. Retrieved May 10, 2018 (https://www.kqed.org/mindshift/51110/when-teens-cyberbully-themselves).

Green, Allison. 2017. "Our 300-Year Love Affair With Advice Columns." *Vox*, June 14. Retrieved June 28, 2017 (https://www.vox.com/first-person/2017/6/14/15782712/advice-columns-dear-abby-history).

Hall, Jeffrey A. and Nancy K. Baym. 2011. "Calling and Texting (Too Much): Mobile Maintenance Expectations, (Over)dependence, Entrapment, and Friendship Satisfaction." *New Media and Society* 14(2):316–331.

Hax, Carolyn. 2018a. "Inattentive Facebooker Misses News of Friend's Tragedy, Commits Faux Pas." *St. Louis Post-Dispatch*, February 24. Retrieved May 15, 2018 (http://www.stltoday.com/lifestyles/relationships-and-special-occasions/columns/carolyn-hax/inattentive-facebooker-misses-news-of-friend-s-tragedy-commits-faux/article_c8d79e7c-a1f3–5f63-aead-f7905316af9e.html).

Hax, Carolyn. 2018b. "Socially Anxious Person Navigates Online Dating." *St. Louis Post-Dispatch*, February 3. Retrieved May 15, 2018 (http://www.stltoday.com/lifestyles/relationships-and-special-occasions/columns/carolyn-hax/socially-anxious-person-navigates-online-dating/article_7fb6dc17-d34e-5681–8418-d791f18982b8.html).

Hax, Carolyn. 2017. "My Niece Made a Nasty Comment About Me on Social Media." *Seattle Pi*, January 12. Retrieved May 15, 2018 (https://www.seattlepi.com/lifestyle/advice/article/Carolyn-Hax-The-hate-is-her-damage-not-yours-10835504.php).

Hertlein, Katherine M. and Amanda Stevenson. 2010. "The Seven 'As' Contributing to Internet-Related Intimacy Problems: A Literature Review." *Cyberpsychology* 4(1):1–8.

Holmes, Linda. 2013. "Dear Reader: Yes There's a Reason People Asked Dear Abby For Advice." *Monkey See: Pop Culture News and Analysis from NPR*, January 17. Retrieved May 10, 2018 (https://www.npr.org/sections/monkeysee/2013/01/17/169627080/dear-reader-yes-theres-a-reason-people-asked-dear-abby-for-advice).

Kent, Robin. 1979. *Aunt Agony Advices: Problem Pages through the Ages*. London: W. H. Allen & Co. Ltd.

Landers, Ann. 1996. "Caught in the Web: The Pros and Cons." *The Chicago Tribune*, November 4. Retrieved May 15, 2018 (http://articles.chicagotribune.com/1996-11-04/features/9611040059_1_internet-addicts-dear-ann-landers-dear-readers).

Lenhart, Amanda and Maeve Duggan. 2014. "Couples, the Internet and Social Media." Pew Research Center, February 11. Retrieved May 15, 2018 (http://www.pewinternet.org/2014/02/11/couples-the-internet-and-social-media/).

Martin, Judith, Nicholaas Ivor Martin and Jacobina Martin. 2018a. "Inviting the World to View a Proposal." *Miss Manners: The Washington Post*, February 20. Retrieved May 10, 2018 (https://www.washingtonpost.com/).

Martin, Judith, Nicholaas Ivor Martin, and Jacobina Martin. 2018b. "Unfriending Etiquette Still Fuzzy." *Miss Manners, Uexpress*, May 2. Retrieved May 10, 2018 (https://www.uexpress.com/miss-manners/2018/5/2/unfriending-etiquette-still-fuzzy).

Melby, Todd. 2010. "Cheating in a Wired World." *Contemporary Sexuality* 44(10):3–10.

Miller-Ott, Aimee. E., Lynne Kelly, and Robert L. Duran. 2012. "The Effects of Cell Phone Usage Rules on Satisfaction in Romantic Relationships." *Communication Quarterly* 60(1):17–34.

Mumsnet. 2014. "Man Texting My Wife." November 21. Retrieved May 15, 2018 (https://www.mumsnet.com/Talk/relationships/2241088-Man-texting-my-Wife-advice-needed).

Nichols, John, ed. [1818] 2012. *The Life and Errors of John Dunton, Citizen of London*. Cambridge, United Kingdom: Cambridge Press.

Patchin, Justin W. and Sameer Hinduja. 2017. "Digital Self-Harm among Adolescents." *Journal of Adolescent Health* 61(6):761–766.

Petronzio, Matt. 2015. "What Makes Someone a Good Advice Columnist?" *Mashable*, June 19. Retrieved May 3, 2018 (https://mashable.com/2015/01/19/advice-columnists/#kdQ5oLnKOsqJ).

Ruble, Traci. 2013. "Is 'Friending' Your Ex on Facebook a Good Idea?" *GoodTherapy.org*, July 17. Retrieved May 7, 2018 (https://www.goodtherapy.org/blog/is-friending-your-ex-on-facebook-a-good-idea-0717134).

Sanchez, Matthew and Quentin Smith. 2011. "'Friending' Personnel Problems? Legal Implications of Social Media in the Workplace." Employment Law Breakfast Program, March 3. Retrieved May 6, 2018 (https://www.sheehansheehan.com/wp-content/uploads/2015/09/personnel_problems.pdf).

Schwartz, Alan. 2013. "Friending and Unfriending on Facebook." *MentalHelp.Net*, February 19. Retrieved May 7, 2018 (https://www.mentalhelp.net/blogs/friending-and-unfriending-on-facebook/).

Sibona, Christopher. 2014. "Facebook Fallout: The Emotional Response to Being Unfriended on Facebook." Conference: Proceedings of the 2014 47th Hawaii International Conference on System Sciences. DOI 10.1109/HICSS.2014.218.

Statista. 2018a. "Percentage of US Middle School and High School Students Who Were Cyberbullied as of August 2016, by Type and Gender." Retrieved May 10, 2018 (https://www.statista.com/statistics/291034/cyber-bullying-share-of-us-students-by-type-and-gender/).

Statista. 2018b. "Reach of Selected Social Networks in the United States as of February 2017, by Age Group." Retrieved May 10, 2018 (https://www.statista.com/statistics/305245/us-social-network-penetration-age-group/).

Statista. 2018c. "Typical Social Media Posts of Adults in the US as of June 2016." Retrieved May 10, 2018 (https://www.statista.com/statistics/616624/us-adults-typical-social-media-posts/).

Trend Micro. 2018. "Public or Private?: The Risks of Posting in Social Networks." Trend Labs Security Intelligence Blogs. Retrieved May 15, 2018 (https://blog.trendmicro.com/trendlabs-security-intelligence/the-risks-of-posting-in-social-networks/).

US Department of Health and Human Services. 2018. "What is Cyberbullying?" stopbullying.gov, February 7. Retrieved May 3, 2018 (https://www.stopbullying.gov/cyberbullying/what-is-it/index.html).

Van Buren, Abigail. 2018a. "Dear Abby." *The Durango Herald*, April 24. Retrieved May 7, 2018 (https://durangoherald.com/articles/220196).

Van Buren, Abigail. 2018b. "For Months, Wife and Her Ex Shared Late Night Call, Texts." *Chicago Sun Times: Lifestyles*, April 4. Retrieved May 7, 2018 (https://chicago.suntimes.com/lifestyles/dear-abby-for-months-wife-and-her-ex-shared-late-night-calls-texts/).

Van Buren, Abigail. 2017a. "Evenings at Home Fail as Wife Embraces Cellphone." *Dear Abby Uexpress*, August 24. Retrieved May 10, 2018 (https://www.uexpress.com/dearabby/2017/8/24/evenings-at-home-fall-silent-as).

Van Buren, Abigail. 2017b. "Girl Sees Red Flags in Her Online Friendship Oversees." *Dear Abby Uexpress*, November 3. Retrieved May 10, 2018 (https://www.uexpress.com/dearabby/2017/11/3/girl-sees-red-flags-in-her).

Van Buren, Abigail. 2017c. "Husband Gripes about Me on Facebook." *Chicago Sun Times: Lifestyles*. December 20. Retrieved May 7, 2018 (https://chicago.suntimes.com/lifestyles/dear-abby-husband-gripes-about-me-on-facebook/).

Van Buren, Abigail. 2017d. "Son Cuts Ties After Parents Block His Texts." *Dear Abby Uexpress*, November 8. Retrieved May 10, 2018 (https://www.uexpress.com/dearabby/2017/11/8/divorced-mom-mulls-overruling-dad-on).

Van Buren, Abigail. 2017e. "Wife's Private Photo Album Makes its Way onto the Web." *Dear Abby Uexpress*, November 29. Retrieved May 10, 2018 (https://www.uexpress.com/dearabby/2017/11/29/wifes-private-photo-album-makes-its).

Van Buren, Abigail. 2017f. "Woman is Ready to Hang Up on Constantly Calling Husband." *Dear Abby Uexpress*, December 5. Retrieved May 10, 2018 (https://www.uexpress.com/dearabby/2017/12/5/woman-is-ready-to-hang-up).

Van Buren, Abigail 2016a. "Friend Ignores Me to Text at Dinner." *Chicago Sun Times*, October 18. Retrieved May 15, 2018 (https://chicago.suntimes.com/lifestyles/dear-abby-friend-ignores-me-to-text-at-dinner/).

Van Buren, Abigail. 2016b. "Wife's Job at the Gym Has Husband in a Sweat." *Aiken Standard*, November 26. Retrieved May 15, 2018 (https://www.aikenstandard.com/wife-s-job-at-the-gym-has-husband-in-a/article_9f298a1c-b5ac-11e6-8419-5b0203451bcc.html).

Van Buren, Abigail. 1981. *The Best of Dear Abby*. Kansas City, MO: Andrews and McNeel.

Vernon, Amy. 2014. "Dear Amy: Sharing My Grandkids Instagram Photos #SMEtiquette." January. Retrieved May 7, 2018 (https://www.amyvernon.net/dear-amy-sharing-my-grandkids-instagram-photos-smetiquette/).

Vilhauer, Jennice. 2015. "This is Why Ghosting Hurts so Much." *Psychology Today*, November 27. Retrieved August 30, 2018 (https://www.psychologytoday.com/us/blog/living-forward/201511/is-why-ghosting-hurts-so-much).

Ward, K. 2006. "The Bald Guy Just Ate an Orange: Domestication, Work and Home." Pp. 145–164 in *Domestication of Media and Technology*, edited by T. Berker, M. Hartmann, Y. Punie, and K. Ward. Berkshire, England: Open University Press.

Part III

VIRTUAL AGENCY AND DIGITAL DYSTOPIA

Preface

These four chapters examine the dynamic relationship among social actors and institutions in the conjoined physical and digital space context. Macro-social factors are emphasized, although both macro-social and micro-social implications are addressed. At one level, conscious versus naïve choices about digital behavior are considered in view of the potential for self-empowerment and collective action. Also addressed is the realization of social agency—the ability for persons to make and act on their own choices—or the diffusion of such agency due to the realities of digital space environments. Ultimately, the reach, and relative effectiveness, of social control through major social institutions, the state, and capitalist enterprise are considered in terms of the potential for positive social change, and in terms of what could become a dystopian digital panopticon.

Chapter 10 addresses music piracy in view of social control and the social construction of deviance—or how different social cohorts have developed particular shared ideas about the meaning of deviance. The chapter starts with a social-historical treatment of law and enterprise that define music piracy as "deviant." One focus is the theoretical treatment of actual music piracy behaviors, versus the active seeking of normative change. The context can be partially framed as a type of *moral entrepreneurship* where digital music consumers, through the prevalence of their behavior, potentially influence the normative outlook on music "piracy." The discussion of the legal, commercial, and social theory divides around music piracy is followed by the discussion of Okamura's survey results around music piracy attitudes and behaviors.

Chapter 11 explicitly ties activism to social network theory, which is unfolded in the chapter. It also addresses macro-social concepts of control in a social-historical look at suppressive political and institutional responses to social movements supported by digital networking. In his theoretical frame, Apardian emphasizes that *social networks* and *social networking* are theoretical constructs that study the *relationships among* individuals, groups, organizations, and societies. He extends this to examine whether digital activist websites fulfill the same social agency function as those indicated by classic networking theory, and what might mitigate that potential. He draws on Tonnies ideas of the social contract in discussing the shift from localized, physical space community, to sets of amorphous, digitally-based communities. Finally, classic social capital theory is explained and addressed through this shift.

Chapter 12 is explicitly theoretical in its treatment of the origins, development, and contemporary ubiquity of the Internet. Choi provides a macro-level theoretical focus on power dynamics, institutional social control, and hegemonic ideology. All of which are unfolded in a straightforward, social-historical explanation of the development of the Internet. The critical approach is conflict theory as manifested in how the interests of the military-industrial, power-elite affected the invention and evolution of the Internet. War-focused ideology and capitalist economic ideology are treated as hegemonic constructs.

Chapter 13 implicitly draws from some of the same conflict theory ideas explicated in Chapter 12. It uses concepts of capitalist ideology and the empowered interests of the commercial elite to address the manipulation of Internet users. Overt, covert, and semi-covert big-data gathering dynamics are explored in terms of coercive power. The subsequent contribution to sophisticated mass manipulation is treated in terms of ideologically-rooted social construction of perceptions. Concrete examples and analogies are used to illustrate these dynamics. Nitzani provides a somewhat polemical concluding set of suggestions about countering some elements that limit social agency during Internet use.

10

ISLANDS IN THE STREAM

How Digital Music Piracy Became a Normal Activity

Daniel Okamura

The Tech and the Law

The ways people buy and listen to music have changed rapidly over the last twenty years, and those changes do not appear to be slowing down. The MP3 format for compressed audio was invented in 1993, and software MP3 players became popular in 1997 with the release of Winamp. Portable hardware players became available in 1998 but did not become popular until 2001 when Apple released the iconic iPod. While it was not the first digital music player without an optical drive, it was the first to become widely successful. However, most music was not available to legally download until 2003, when Apple opened the iTunes music store. Both software and hardware MP3 players required us to either extract songs from CDs and convert them into MP3 files, or to download such files from other people—both of which were, and continue to be, illegal in most countries. To many people, there was no reason why the average person would even bother to wonder whether a program or a device they legally obtained to play music would ask them to engage in illegal acts in order to use them. These factors still apply today.

In the United States, the Digital Millennium Copyright Act (DMCA) prohibits nearly all copies made from CDs, no matter what we wish to use them for. When the law was written in 1996, media companies assumed that piracy would mainly occur in the form of selling bootlegged copies of albums and movies, and their trade groups—the Recording Industry Association of America (RIAA) and the Motion Picture Association of America (MPAA)—lobbied Congress to craft the law accordingly. The primary focuses of the law are on making money and preventing copying information from optical discs such as CDs and DVDs.

The resulting system of piracy involves three groups with different goals: (1) media corporations, (2) the legal system, and (3) consumers. The media companies are primarily driven by the motive to make profits and protect their intellectual property using the legal system. Lawmakers craft regulations under the influence of trade groups like the RIAA, while judges and attorneys uphold the law and provide the means for the RIAA to take civil action against copyright infringers. Finally, consumers both purchase and steal music—a combination of behaviors that helps determine the profits made by media companies—but also find easier ways of obtaining music when legal means of doing so are more difficult than illegal ones.

As technology changes, so does our access to music. Once our networking capabilities and compression technologies could transfer music more rapidly than a CD player, the CD's importance began to dwindle. However, without a physical disc to trade hands, categorizing music becomes problematic. Consumers, corporations, academic researchers, and legal entities do not agree on whether we should consider music part of "goods" or a "service" (Helberger et al. 2013). While a media company might refer to its "products" or its "holdings," the United Nations Educational, Scientific, and Cultural Organization (UNESCO) recommends that digital files be treated as cultural expressions rather than consumer goods. This suggests a poor match between digital consumption and the laws that govern it. For example, media companies claim that, by using their products, consumers are consenting to the terms of their digital rights management (DRM) software. However, DRM tends to unduly limit digital books, movies, and music more than their physical counterparts (Helberger et al. 2013). Behaviors we take for granted in the modernist age now face new challenges. How do we lend someone a book or an album we think they will enjoy? How do we sell—or give away—the ones we no longer want?

What Does Piracy Really Mean?

Researching digital music consumption—and especially piracy—has proven difficult because researchers, corporations, legal structures, and consumers all have different definitions of piracy. Hsieh and Lee (2012:364) expressed such difficulty in comparing their study on "unauthorized software copying" to previous ones, because of "different descriptions of dependent variables made and different numbers or combinations of factors proposed." It no longer seems sufficient to express music piracy as "the activity involving illegal reproduction (or acquisition through illegal means) of digital sound without an explicit permission from the copyright holder" (Nandedkar and Midha 2012:41). What "activity" does that mean? What is legal and illegal? How do we know if we have permission from the copyright holder? Who is the copyright holder? Under the DMCA in the US, it is possible to understand piracy to include (1) giving and taking digital music without paying the copyright holder, (2) extraction of music data from CDs, and (3) the duplication of CDs.

This chapter is primarily concerned with the behaviors involved in pirating music. Many behaviors that people already engage in may be associated with piracy, because music, by its very nature, is a social phenomenon. Thus, piracy may be embedded in very common acts, like listening to music with friends or making a mixtape for our significant others (Cammaerts 2011).

Economics—Product? Creator? Knowledge?

To understand what is at stake, we should return to the definition of *copyright*. In a critical *New Yorker* essay, Lewis Menand gets right to the point: "A copyright is, first and foremost, the right to make a copy" (2014). A consumer who purchases a copyrighted book can do nearly anything to it—use it to hold up a wobbly table, run it through a paper shredder to make confetti, or give it away. The one thing the book purchaser cannot do, is make a copy of it. A consumer who were to make unauthorized copies of the book, could make money selling those copies without spending the time and money it took to write the book in the first place. The *product* is central to this definition. On the other hand, Waldfogel (2012) describes "copyright" as a government-issued monopoly on a product, as an incentive for the innovator of that product to make more products so supply of creative products never dwindles. The *creator* is central to this definition. While American and European copyright philosophy differs, both are there to let authors make a profit for a limited time, which gives them time to work on new ideas. This, in turn, ensures supply of new works (Handke 2012).

Cammaerts (2011) observes that copyright is an idea of the industrial age (as opposed to the postmodern era) because it implies either *possessing an object*, or restricting access to *be near to an expression*. Menand (2014) suggests, however, that this is not about technology. Instead, he points to Article I of the US Constitution, wherein Congress is trusted "to promote the Progress of Science and useful Arts." This perspective supports the idea that, by rewarding authors for their writing and, consequently, encouraging them to produce more creative and scientific works, copyright protection exists to accelerate the spread of knowledge.

Analogous arguments about the effect of digital piracy have been put forward. On closer examination, however, there are essential differences among them. The economic claim against piracy focuses solely on harming supply by providing less incentive to produce new work. Siwek (2007) observes a $12.5 billion annual loss in the United States and concludes that decreased revenue comes from decreased demand. His paper, however, contains a problem that is typical of this economic approach: it equates one illegal download with one lost sale. This approach overlooks the fact that not everyone who steals a song would have purchased it, and not everyone "gifted" with a pirated song would have purchased it. Such downloads should, nonetheless, be included in the definition of demand for consuming music, but we cannot measure for certain how much this would have translated into lost sales. Despite the inability to explain this in terms of specific monetary amounts this means that demand is up and revenues are down. Piolatto and Schuett (2012) attempt to distill the effect of music piracy into an economic formula and find that popular artists benefit while independent acts lose out. Their paper contains a similar problem typical of this economic approach. They count CD *albums* as the primary unit of sales, despite the upsurge in consumption of *single songs*, both by digital download and streaming—an upsurge recognized by the market in, for example, the ability to legally purchase individual songs from the iTunes Music Store. Both Handke (2012) and Waldfogel (2012) find, however, that digital piracy has not slowed the production of new music, despite falling revenues.

Corporate Legality

If copyright is designed to hasten the free flow of information to the betterment of society, one could actually argue that American legislature and case law have worked to dumb us down. As Menand (2014) observes, the original limit on copyright in 1790 was fourteen years, and you could apply to extend it by another fourteen. He notes, however, that the law changed in 1831 and made copyright extendible to forty-two years after publication. In 1909, that extension was extended to fifty-six years. In 1976, copyright automatically applied for fifty years after the author's death and registering copyrights was no longer required. This made literally everything produced since 1976 subject to copyright protection. In 1998, the fifty-year protection after the author's death was extended to seventy years. Corporate authors were protected for 120 years after creation or ninety-five years after publication. At this time, works published between 1923 and 1978 became retroactively protected for ninety-five years after they were published. A few years prior, in 1993, the European Union even revived outdated copyrights for authors who had died between 1925 and 1944 so their estates could be paid for sales lost during World War II. Clearly, the laws are no longer about providing incentives for creating new works, but ensuring that the copyright holder gets paid for as long as possible.

To this end, Rohter (2011) observes that media companies regularly require musicians to sign away all rights to their recordings. This makes the *corporations* the authors of these recordings, *not the artists*. It does not have to be this way, either, as writers routinely grant corporations one-time publishing rights, and keep the copyright thereafter (Menand 2014).

When a large US record label feels its copyright is being violated, the RIAA has—since 2003—regularly taken legal action against groups and individuals for perceived copyright infringement (RIAAa nd; RIAAb nd). This usually takes the form of a civil lawsuit or even just the threat of one. However, *Ars Technica* writer, Eric Bangermann (2007), observed that Sony

BMG's head of litigation testified that the legal costs of litigating copyright infringers were greater than the damages caused by lost sales. In fact, that Sony executive admitted that RIAA members had not calculated the actual amount of incurred damages. The head counsel for the RIAA later stated in no uncertain terms: "Our goal is to send a message that copyright infringement is wrong and get some compensation for the infringement" (Green 2008). This has now been shown in the RIAA's more recent strategy of sending university students threats to settle for as little as $10 or risk going to trial for $250,000 if they are suspected of copyright infringement (Mitchell 2014). In the case of ReDigi, the desired outcome is less clear. ReDigi was a service that allowed users to sell unwanted music files much like we sell used CDs, records, and books. The RIAA sued ReDigi in 2012 and won, forcing ReDigi offline. The RIAA argued that uploading a "used" music file to ReDigi's servers constituted making an unauthorized copy—even if the source file was to be immediately deleted afterward (Davis 2013). This is to say nothing of streaming services—explicit radio-like services like SoundCloud and Spotify, but also video services like YouTube—whereby consumers listen to music without purchasing it. *Music Business World* finds that, after taxes and the fees each service charges to carry artists, major record labels keep almost three-quarters (73%) of streaming revenues, while writers and publishers get far less (16%), and artists themselves get least of all (11%). This broke down in a French analysis to major labels making €4.56 out of every €9.99 monthly subscription, with the artists themselves taking home just €0.68 (Ingham 2015).

Deviance Theory (or Not) to Predict and Prevent

Most of what little research exists on piracy is firmly situated in theories of deviance. After all, piracy is a crime, so studying it in terms of deviant behavior seems appropriate.

The predictive value of any one particular approach, however, seems less apparent, and agreement among theorists seems poor. Is it that people find ways to morally justify sharing music? Sharing music is not really stealing, right? Do people struggle to choose between social norms they see as important, and friends who they think will enjoy a particular album? Can they simply not control themselves when they want access to a good song? Or is it that people simply think they can get away with music piracy, or that punishment seems vague and unlikely? Or are they simply unmoved by the artist, or the creative value of the music? As discussed below, these are the questions informing much of the research on music piracy.

One argument is that *moral justification* (the ability to rationalize choices that contradict societal rules) predicts music piracy. Further, that such moral justification is stronger than *deficient self-regulation* in predicting illegal music downloading (LaRose and Kim 2007). In these terms, moral justification (how a downloader disengages from the supposed societal norm that piracy is wrong), explains this deviant behavior to a greater degree than deficient self-regulation (how little self-control the person has to resist—even when feeling that piracy is wrong and feeling self-condemnation for doing it).

This is called into question by the observation that people perceive theft and piracy as *different* behaviors and, thus, find that they have different ethical predispositions about them (Wingrove, Korpas, and Weisz 2011; Gopal et al. 2004). In other words, no matter how strongly they may agree that stealing is wrong, this does not factor into their decisions about illegal downloading, because they do not see this type of music sharing as theft.

Researchers also looked at whether resisting acts of piracy might be a matter of how strongly a person believes that conforming to social norms, is an important aspect of the "do no harm" value. Bateman, Valentine, and Rittenberg (2012) identified this as a significant influence on whether a person thinks piracy is immoral, especially in the face of peer pressure. Low self-control has also been put forward as a key factor in whether a person pirates music or not (Higgins et al. 2012). This appears at odds with the position taken by LaRose and Kim (2007) that deficient self-regulation was a relatively unimportant factor. However, it might be that low self-control is important, just not to the same degree as moral justification, when it comes to predicting acts of digital music piracy.

Instead of predicting piracy, some theorists focus on preventing it. Deterring deviance through punishment is a familiar approach (Hinduja and Higgins 2011; Tittle 1980). The claim is that illegal music downloading can be decreased if laws defining piracy and its penalties are made well known. The idea that crime can be prevented if punishment is "swift, certain, and severe" echoes a common trope from more than two and a half centuries ago. However, it misrepresents the original idea discussed in *An Essay on Crimes and Punishments* (Becarria [1764] 2011). Becarria says that the greater the certainty that a crime will be punished, and the more immediately punishment follows the commission of a crime, the more effective the punishment will be as a deterrent. However, he makes it clear that an effective punishment should only be severe enough to make the initial crime not worth committing—as in the familiar phrase, "let the punishment fit the crime." In fact, he points out that the more severe a punishment is, the greater the chance the perpetrator will commit more crimes just to avoid capture and punishment. Consciousness-raising is another familiar approach to changing behavior. Redondo and Charron (2013) discovered that those who never pay for downloads devalue artistry. Findings suggest that the only way to influence them to pay would be to launch a marketing campaign that emphasizes artistic value and plays upon subsequent feelings of guilt.

Still other research avoids deviance altogether and observes the *processes* involved in digital music piracy. Of particular interest is the identification of two different types of pirates, "a smaller group of 'core uploaders' use anonymity services to a greater extent than the average file user" (Larsson et al. 2012:273). These elites are distinguished by more advanced or "professional" practices than someone who occasionally downloads a song. Sintas, Zerva, and García-Alvarez (2012) find that downloading often represents the "path of least resistance" to obtaining music—that sometimes it is easier to download a song without paying for it than to do so using legal channels. Sintas and colleagues (2012) also find that reciprocity online is not determined by the strength of ties in the network. Effectively, this means that when we trade music online with strangers, we are engaging in the same activity and feel no more or less pressure to give back than when trading music with our friends and family. Further, how we perceive the morality of formal (music stores and legal downloading services) and informal music markets (bootleggers and illegal sharing) is related to whether resources are abundant or scarce in such an exchange (Sintas, Zerva, and García-Alvarez 2012). Scarcity results when media companies restrict our access to music instead of permitting us to consume it, which leads us to perceive media companies unfavorably, and possibly, as even owing us something. Perhaps, most importantly, in a CD market, original items are culturally marked as tangible mass-produced discs with logos, labels, booklets, and paraphernalia (Sintas, Zerva, and García-Alvarez 2012). As the technology to exchange music dematerializes it, the distinctions between original and copy drop away, and we are left only with our listening devices to culturally mark music—often in the form of iPhone versus Android.

The Current Music Piracy Study

When a pilot phase of this study was conducted, the questions focused on the flexibility of moral beliefs. However, such questions proved to be too personal, and several respondents quit the survey at the exact same question. Thus, the survey was changed to focus on behavior, asking the 545 participants how they listened to and shared music with their peers: what devices they used, how long and how often they listened to music, what they did while listening to music, and how they both give and receive music with their peers. This approach was taken to try to avoid asking questions perceived as too personal—a circumstance which can lead survey respondents to lie in ways that they think will make them look good to whomever is asking the questions. Deception allows people to avoid sharing what they think might be an unpopular opinion or activity. Deceptive responses have been observed even when asking questions about age, race, income, and education.

Participants

Most of the respondents to this survey were young, white, cisgender men. Slightly more than half (55%) of the 545 people who took the survey were between 18 and 34 years of age, the so-called "Millennial Generation," and nearly all respondents (94%) were under 50 years of age. White people made up more than three-quarters of the sample (78%) and men made up more than half (57%).

What Do They Listen To, and When?

People listen to music in a variety of ways—most individuals use multiple devices to access music. However, not all means are equally popular. Most survey respondents said they listened to music on their computers. This may be true partly because they listen while simultaneously engaged in other computer use. Similarly, a large percentage consume music through YouTube, closely followed by the number who use a streaming service, which is immediately followed by smartphone music users. Radio completes the list of high-use devices. These are the current "winners" in the music business, as reflected in being chosen by almost three-quarters or more of the respondents (see Table 10.1). Given that more and more younger Millennials are starting to access the Internet only from smartphones, one would predict that the smartphone will move up in the list before long.

There is a significant drop in the proportion of respondents who use other means (see Table 10.2). The less used music sources included dedicated MP3 players, compact discs, tablet computers, music links that respondents "discovered," links sent to them, YouTube-to-MP3 converters, and musicians' websites.

As will be shown later in the discussion, however, Table 10.2 data is contradicted by answers to questions about giving and receiving links—which show links to be quite frequently given and slightly more frequently received (each more than twice the proportion reported in Table 10.2). Surprisingly for this author, only 58 percent of the sample reported listening to files they *purchased* to download.

When asked what they do while listening to music, most people said they were commuting (89%) or engaged in other dull tasks like housework (68%), office work (62%), and exercising (59%). These answers appear to be

Table 10.1 Most Popular Music Sources

Personal computers	85%
YouTube	77%
Streaming services	76%
Mobile phones	75%
Radio	71%
Average percentage who access these devices	77%

Table 10.2 Less Popular Music Sources

MP3 players	43%
Compact discs	38%
Tablet Computers	29%
Found music links	28%
Shared/sent music links	25%
YouTube-to-MP3 converters	24%
Musician websites	22%
Average percentage who access these devices	30%

supported by the duration and frequency of listening patterns: Almost two-thirds (72%) listen to music almost daily, although only about one in five (22%) listen for more than four hours at a time.

Music Sharing

If people are not listening to music using CDs, or links they find or receive from friends, and just over half are listening to music they bought, where are they getting their music? The responses to how people share music with others may hold the answer. People shared music with their friends and family by sending them links (64%) and sending them files (52%). When asked how they received music, slightly more than half (54%) received burned CDs, but the receiving pattern largely matched the giving pattern: most received links (69%) or files (55%). These results directly contradict the data from the question asking how people listened to music. Respondents said they used their phones to listen to music, but only about one-quarter said they got their music from links they found or links others sent to them. While three-quarters of the sample streamed their music, the rising popularity of streaming services does not account for only half of those who said they shared links and files actually listening to them. Clearly, respondents still felt uncomfortable answering honestly, even when talking indirectly about music piracy.

When asked for their attitudes about music sharing, many people strongly agreed that sharing music increases bonding (45%), sharing via download is common (56%), and that downloading is common (56%). Large portions of the sample also condemned piracy, answering that downloading for personal use hurts musicians (30%), and downloading to share hurts musicians (35%). Interestingly, just over one-third (36%) of respondents preferred having files to streaming, suggesting that piracy is not going away as streaming gains popularity—people still want to "own" something.

Perceived Legality

Participants were also asked about what they thought was legal where they were. Many said they did not know whether it was legal to back up their music (43%), but many thought it was (33%). Similarly, almost one-third (31%) of the sample did not know whether it was legal to copy their CDs onto an MP3 player, but nearly half (48%) thought it was. Only a few nations with small music markets allow for making backup copies of music or for extracting songs from CDs to put on your iPod at this time. It follows then, that, regardless of whether it is legal or not, most (86%) of those surveyed thought it should be legal to make backup copies of their music, and about two-thirds (65%) thought it should be legal to rip songs to use on an MP3 player.

Reciprocity

Data were also analyzed to look for correlations between variables. The strongest correlations occurred between pairs of giving and receiving behaviors, suggesting that people consider giving music to be part of the same family of behaviors as receiving music. Participants were asked whether they give people music gift cards, new compact discs, used compact discs, burned compact discs, links to artists' websites, links to musical content (including pirated files), and actual files. They were also asked whether they received the same things from others. Each giving activity highly correlated with its receiving counterpart.

Even when looking at giving behaviors and receiving behaviors by themselves, the strongest correlations suggested that respondents who said they participated in one behavior were likely to participate in most of them. This means that if a person said they shared music with their friends and family by giving them a new CD, they were also likely to give people music gift cards, used CDs, burned CDs, and MP3 files. Receiving behaviors showed a similar effect: those who said they received new CDs from others were also likely to receive music gift cards, used CDs, and burned CDs.

Oddly, there were no negative relationships between people who listened to music via legitimate means (such as the radio or compact discs), and those who pirated via files and links—in other words, a higher proportion of legitimate consumption did not predict lower illegitimate consumption. There were also no strong associations between attitudes and behaviors. Most people who supported piracy outright agreed with most statements about it—except they believed that illegal downloading harms musicians.

Even though the sample was heavily imbalanced toward young, white, cisgender men, demographic questions did not yield much in terms of strong correlations. Respondents aged 18 to 34 were only slightly less likely than older adults to listen to compact discs, give their friends gift cards for music, or receive music gift cards from their friends. Cisgender men were slightly less likely to listen to the radio than cisgender women. Race and ethnicity yielded no significant effects, which is interesting considering that non-White people in the US are up to three times as likely to only be able to access the Internet via their smartphone. However, this could simply be an indicator that a lot of people took the survey using a personal computer rather than a phone.

Discussion

Within this sample of 545 people, more than two-thirds said they give, and about the same number say they receive, music links. Similarly, over half report said they give music files, with a comparable number saying they receive music files. Apparently, the most common ways for respondents to obtain digital music is through links and files, which contradicts the sample's moderately strong condemnation of piracy. This also contradicts the lack of correlations between attitudes and actual behavior, suggesting people may not have been entirely truthful in their answers to this survey. This may inform future research in that respondents may lie, not only when discussing socially-unfavorable behavior, but when there is any possibility that the behavior they are discussing may make them look bad.

Respondents said that piracy is a common activity, which supports claims by LaRose and Kim (2007) that moral justification helps determine whether a person will pirate music or not—after all, everyone else is doing it. Such justification and disagreement with piracy laws suggest that this behavior is well-established and unlikely to change soon, despite the adoption of formal digital music marketplaces. Those who took this survey did not know whether their behaviors were legal, but overwhelmingly believed they should be.

We cannot blame the consumer or underground movements for the normalization of the use and piracy of digital music, either. Digital music players have been widespread as software since 1997, as portable devices for purchase since 1998, and their popularity rapidly expanded after the release of the iPod in 2001. With no legal means of populating them with music, these programs and devices effectively commodified and legitimized piracy behavior to mainstream markets. Further, legal digital marketplaces were not available to the consumer for six years after the first programs became popular, and five years after portable digital music players hit the market, giving these behaviors ample time to coalesce into norms. Moreover, as the CD continued its decline in significance, consumers responded by questioning why they would want to purchase a plastic disc they had no intention of using.

This is where the economic claim against piracy breaks down. Existing copyright philosophy combines incentive to the creator with an expectation of more and faster spread of products, accelerating the diffusion of information and culture. However, the laws justified by such philosophy only provide for payment to the copyright holder, and do not provide for the creator. Previous research demonstrates that, while revenue has decreased, supply of and demand for music has increased. The results of this survey support those claims. In addition, results show that the attitudes and behaviors considered normal by consumers imply that consumption of music has increased, as well. The music industry has understandably regretted the loss of CDs, which always contained more than one song, allowing record companies to justify an increased price point. The switch to digital music has meant that consumers more often consume the discrete items they want, without feeling like they are being stuck with unwanted material. In strictly economic terms, this means the supply-demand chain has become more efficient, and it remains incumbent on record companies to more effectively monetize consumption.

References

Bangermann, Eric. 2007. "RIAA anti-P2P Campaign a Real Money Pit, According to Testimony." Arstechnica.com, October. Retrieved September 15, 2014 (http://arstechnica.com/tech-policy/2007/10/music-industry-exec-p2p-litigation-is-a-money-pit/).

Bateman, Connie R., Sean Valentine, and Terri Rittenburg. 2012. "Ethical Decision Making in a Peer-to-Peer File Sharing Situation: The Role of Moral Absolutes and Social Consensus." *Journal of Business Ethics* 115(2):229–240.

Becarria, Cesare. [1764] 2011. *An Essay on Crimes and Punishments*. Indianapolis: Online Library of Liberty. Retrieved September 22, 2015 (http://oll.libertyfund.org).

Cammaerts, Bart. 2011. "The Hegemonic Copyright Regime vs. the Sharing Copyright Users of Music?" *Media, Culture & Society* 33(3):491–502.

Davis, Wendy N. 2013. "Good Copy, Bad Copy." *American Bar Association Journal* 99(8):14–15, 18.

Gopal, Ram D., G. Lawrence Sanders, Sudip Bhattacharjee, Manish K. Agrawal, and Suzanne C. Wagner. 2004. "A Behavioral Model of Digital Music Piracy." *Journal of Organizational Computing and Electronic Commerce* 14(2):89–105.

Green, Heather. 2008. "Does She Look Like A Music Pirate?" *Businessweek.com*, April 23. Retrieved September 14, 2014 (http://www.businessweek.com/stories/2008-04-23/does-she-look-like-a-music-pirate).

Handke, Christian. 2012. "Digital Copying and the Supply of Sound Recordings." *Information Economics and Policy* 24(1):15–29.

Helberger, N., Marco B.M. Loos, Lucie Guibault, Chantal Mak, and Lodewijk Pessers. 2013. "Digital Content Contracts for Consumers." *Journal of Consumer Policy* 36(1):37–57.

Higgins, George E., Catherine D. Marcum, Tina L. Freiburger, and Melissa L. Ricketts. 2012. "Examining the Role of Peer Influence and Self-Control on Downloading Behavior." *Deviant Behavior* 33(5):412–423.

Hinduja, Sameer, and George E. Higgins. 2011. "Trends and Patterns Among Music Pirates." *Deviant Behavior* 32(7):563–588.

Hsieh, Pei-Hsuan and Tze-Kuang Lee. 2012. "Does Age Matter? Students' Perspectives of Unauthorized Software Copying under Legal and Ethical Considerations." *Asia Pacific Management Review* 17(4):361–377.

Ingham, Tim. 2015. "Major Labels Keep 73% of Spotify Premium Payouts — Report." Music Business Worldwide, February 3. Retrieved October 20, 2015 (http://www.musicbusinessworldwide.com/artists-get-7-of-streaming-cash-labels-take-46/).

LaRose, Robert, and Junghyun Kim. 2007. "Share, Steal, or Buy? A Social Cognitive Perspective of Music Downloading." *CyberPsychology and Behavior* 10(2):267–277.

Larsson, Stefan, Måns Svensson, Marcin de Kaminski, Kari Rönkkö, and Johanna A. Olsson. 2012. "Laws, Norms, Piracy and Online Anonymity: Practices of De-identification in the Global File Sharing Community." *Journal of Research in Marketing* 6(4):260–280.

Menand, Louis. 2014. "Crooner in Rights Spat: Are Copyright Laws Too Strict?" *The New Yorker*, October 20. Retrieved October 13, 2015 (http://www.newyorker.com/magazine/2014/10/20/).

Mitchell, Jennifer. 2014. "Music Industry Targets UMaine Students in Piracy Crackdown." The Maine Public Broadcasting Network. Retrieved October 20, 2015 (http://www.mpbn.net/Home/tabid/36/ctl/ViewItem/mid/5347/ItemId/32390/Default.aspx).

Nandedkar, Ankur, and Vishal Midha. 2012. "It Won't Happen to Me: An Assessment of Optimism Bias in Music Piracy." *Computers in Human Behavior* 28(1):41–48.

Piolatto, Amedeo, and Florian Schuett. 2012. "Music Piracy: A Case of 'The Rich Get Richer and the Poor Get Poorer'." *Information Economics and Policy* 24(1):30–39.

Redondo, Ignacio, and Jean-Philippe Charron. 2013. "The Payment Dilemma in Movie and Music Downloads: An Explanation Through Cognitive Dissonance Theory." *Computers in Human Behavior* 29(5):2037–2046.

RIAAa (Recording Industry Association of America). nd. "About Piracy." Retrieved September 10, 2018 (https://www.riaa.com/resources-learning/about-piracy/).

RIAAb (Recording Industry Association of America). nd. Why We Do What We Do." Retrieved September 10, 2018 (http://www.riaa.com/physicalpiracy.php?content_selector=piracy-online-why-we-do-what-we-do).

Rohter, Larry. 2011. "Record Industry Braces for Artists' Battles Over Song Rights." *Nytimes.com*, August 16. Retrieved September 16, 2014 (http://www.nytimes.com/2011/08/16/arts/music/springsteen-and-others-soon-eligible-to-recover-song-rights.html?_r=0&pagewanted=print).

Sintas, Jordi López, Konstantina Zerva, and Ercilia García-Alvarez. 2012. "Accessing Recorded Music: Interpreting a Contemporary Social Exchange System." *Acta Sociologica* 55(2):179–194.

Siwek, Stephen E. 2007. "The True Cost of Sound Recording Piracy to the US Economy." Policy Report 188. Institute for Policy Innovation, October 10. Retrieved September 11, 2018 (https://www.ipi.org/ipi_issues/detail/the-true-cost-of-copyright-industry-piracy-to-the-us-economy).

Tittle, Charles R. 1980. *Sanctions and Social Deviance: The Question of Deterrence*. New York: Praeger.

Waldfogel, Joel. 2012. "Digitization and Copyright: Some Recent Evidence from Music." *Communications of the Association for Computer Machinery* 55(5):35–37.

Wingrove, Twila, Angela L. Korpas, and Victoria Weisz. 2011. "Why Were Millions of People Not Obeying the Law? Motivational Influences on Non-Compliance with the Law in the Case of Music Piracy." *Psychology, Crime and Law* 17 (3):261–276.

11

COMMUNITY OR CATHARSIS?

Online Activism, Digital Community, and Social Agency

Berge Apardian

Activism and Social Networks

Activism has significantly changed in recent decades. With the advent of the digital age, activism has become a multi-layered process that takes place in both physical and digital space. This affects the way it is facilitated, and how it is felt. Promoting activism through one's physical presence (protesting, rallying, etc.), is experienced differently and has different impacts when compared to engaging activism through one's digital presence. One issue is whether individuals are more likely to participate in activism due to social-cause-related websites on the Internet. A second, even more salient issue, is whether or not online activism has translated into material actions in social movements or behaviors that contribute to community-engaged activism. In other words, how effective is the activism involving one's digital presence? This chapter addresses concerns around digital media's immediate and overall impact on activism, concrete social movements, and more localized community engagement. The insights offered throughout are based, in part, on an analysis of user interactivity across eighty-five activism-dedicated websites.

Activism consists of efforts to change the social, economic, political, and/or environmental status quo. It is comprised of a social movement—or several social movements—that include persistent and organized efforts on the part of a relatively large number of people, either to bring about, or to resist social change (DeFronzo 2011:15).

Activism requires a social network. In the sociological sense, *social network* and *social networking* are theoretical constructs that study the relationships among individuals, groups, organizations, and societies. A network is recognized in the creation and expansion of one's social contacts, as one makes connections through individuals and groups (Rouse 2006). Successful activism, therefore, requires an effective utilization of social networking sites (SNSs), and the creation of a social network through them.

The Internet promotes such connections through web-based platforms, which are known as *social networking sites.* These interest-based sites may be dedicated to multiple social causes or a single social cause; they may only incidentally or sporadically refer to social causes that may be trending; or they may have no connection at all to social causes. This chapter is concerned with activism and the Internet, generally speaking, as well as websites that are specifically dedicated to activism.

Internet social networking sites establish digital communities, which ultimately help people make and keep contacts with others (Rouse 2006). This can indeed elongate one's digital presence. Strengthening one's sense of digital presence may seem to support activism, and under some circumstances it does. Generally, however, to be effective, website-supported digital activism still depends on accompanying material action through physical presence. Catharsis, and a false sense of social agency, may decrease effective activism by making potential activists feel like they have already taken sufficient action when they respond to cause-related Internet sites. The context created by complex relationships among technology, society and human emotion has been recognized as a salient concern (Zilberstein 2015). Internet sites also have the potential to decrease activism due to the possibility of feeling overwhelmed by the multiplicity of Internet-based "calls for action." Not being able to address every cause that inspires serious concern can be deeply frustrating. Activism is generally motivated by strongly felt responses to particular conditions. Catharsis provides relief through venting or addressing strong emotions. It is important to explore whether the activism-focused, websites are more likely to provide catharsis than they are to promote effective activism.

Historically, digital media has provoked organized action in the physical world, as well as providing a digital environment for elevated social agency (the power to act without excessive social constraint), but now it seems possible that SNSs can dissipate activism as well. Before moving on to a discussion of how activism plays out in digital and physical space today, it is important to consider high visibility examples from the recent past, and look at why these may or may not still be viable models for digitally supported, significant social change.

Activism and Digital Space

Digital media has created and expanded various social networking venues which were utilized and responsible for the revolutionary movements in the Middle East, often referred to as the "Arab Spring." In the year 2010, Tunisia began a revolution that, in 2011, ultimately resulted in the ousting of the long-time president Zine El Abidine Ben Ali (who had reigned as totalitarian ruler for twenty-three years). This revolution triumphed due to the utilization of social networking sites such as YouTube, Facebook, and others. Following the Tunisian revolution, Egypt began to undergo a similar process which ultimately led to the ousting of President Hosni Mubarak (who had ruled for thirty years). The use of social media sites in Tunisia and Egypt improved acts of activism in four ways.

First, it made it easier for disaffected citizens to act publicly in coordination. This is a classic example of social agency. Second, by creating information cascades about the likelihood for success of the revolt, there was widespread confidence among the people. Third, rulers began paying a higher cost for the continued repression of their people (Youmans and York 2012). Lastly, diffusion of information to regional and global publics became possible, while dramatically increasing publicity around the world. This process made the events in the revolution public to the world, thus crippling the regime by making any attempts to fabricate the truth nearly impossible.

During the revolt in Egypt however, Mubarak underwent a different attempt to battle the uprising. First, he blocked Facebook and Twitter; but protestors rapidly found alternate digital platforms though which to congregate. Attempting to stop the use of social media networks completely, Mubarak then turned off all Internet satellites, digitally isolating Egypt from the rest of the world. Such efforts were futile—coming too late after initial social networking among activists who were then prepared to resort to more direct methods of activism, such as taking over the streets of Egypt and congregating physically (Youmans and York 2012). This worked.

Knowing how social networking sites were effectively utilized for collective action in Tunisia and parts of the Egyptian revolt, people in Libya, Syria, Iran, and Sudan began to attempt similar organizing for social change. Libyans were successful in their revolution; however, the leaders of Iran, Syria, and Sudan became aware of the surveillance possibilities using social networking sites, and were consequently able to use those platforms for gathering information on any potential plans for activism. They were able to turn digital media use around, as an anti-revolutionary weapon against activists. This led governments to invest time and money into the creation of an "electronic army," where (through digital media and the Internet) all civilians could be kept under surveillance (Youmans and York 2012). These types of public policies ultimately cripple digital activism in these environments because participation moves from anonymity and selective visibility, to vulnerability and exposure. Activists have their own organizing platforms used against them. Even more disastrous for activism, the power of oppressive social agencies are able to extend their reach using digital means of gaining access to the lives of their citizenry—disregarding privacy—in the name of national security.

If this is the background from which to consider digital space activism in the present, what is the background for considering local community-engaged, "feet on the ground" activism in the present US? Perhaps the most important issue to address is the decentralization of community and the accompanying shift from socially rooted to instrumentally-motivated relationships.

Loss of Local Community and Activism in Physical Space

Tonnies ([1887] 2017) described the transformation from traditional neighborhoods to modern cities as a major contributor towards the loss of community, and the rise of *individualistic* tendencies—where self-interest dominates communal interests. Cities replaced traditional communities. During the transformation, social links once based on trust and geographic proximity became replaced by social ties based on instrumental (pragmatic) rationale and profit. Tonnies referred to such instrumental ties as the *social contract.*

As traditional societies grew towards more modern urban grids, and individuals began to commute further for work, people began spending more time away from home and the community in which they resided. Traditionally, working with families in a neighborhood would have increased the chances of the neighborhood thriving, ensuring the continuity of local stores where families could work. The money people made would continue to circulate within the town. For example, if the baker bought flowers from the florist for five dollars, the florist used the same five dollars to purchase fish from the local market owner, who then bought apples from the local store, who then used the same five dollars to purchase bread from the baker. Thus, the baker receives his/her money back. Not only would there have been a direct tie to their survival, but also to one another. This was the source of social capital as well—social capital is formed by the social relationships among people who live and work in close proximity.

Instead, people began to spend more time with secondary contacts rather than with their primary and more intimate counterparts. Individuals no longer worked with members from their communities, but with people who lived in different areas altogether, simply to meet at a work site. The result was a shift in work relationships. This process causes individuals to form social ties and social networks with those who will serve a rational and material or profitable purpose. Trustworthiness and personality-driven qualities are no longer necessary to carry out day-to-day activities and business relations. People no longer depend on long-lasting work relationships. Ultimately, they no longer need to feel a close community-based relationship with co-workers (Wirth [1938] 2010).

Consequently, as people spend much more time with business associates than with neighbors, townsfolk, and family, they lose a sense of communal bonds. These are replaced by relationships that aim to satisfy individual, rather than shared community needs. This potentially disconnects individuals from their neighborhoods and causes a decline in local activism, because it focuses the attention on individualistic desires that are based on profit accumulation and entertainment. It takes away the need to spend more time with locals in order to congregate and discuss things that are happening in their "own backyard." It diminishes the overall knowledge of what is happening locally, as well as the overall time available to spend participating in activism. This process has continued and escalated since first described by Tonnies over a century ago—so what happens to activism? It has been argued that the Internet has played a role in this as well. That the Internet has accelerated the "transformation in community from densely knit villages and neighborhoods to more sparsely knit social networks" (Rainee et al. 2006).

Back to Digital Connectivity

At work, people tend not build deep and abiding complex social relationships with commuting co-workers. They use their smartphones, and rely on their digital presence, for a sense of connectivity or a community—even if it is a relatively-narrow community of interests, rather than broadly-interconnected physical community of shared local, social, political, economic, and regional life. With primary contacts, it enables a sense of belonging to something more intimate (i.e. seeing pictures and posts from family), while being somewhere less intimate (work). This acts to reduce the perceived emotional distance from one another. When Wirth predicted that a population of commuters would become alienated from intimates due to urban sprawl, he could not have anticipated digital connectivity. The author argues, however, that this is a double edge sword. While digital connectivity makes the added commute-time separation from loved ones bearable, it further ensures the continuation of urban sprawl. Nevertheless, it counterbalances the depth of alienation and instrumentality predicted by early urban social theorists.

Social networks did not entirely dissipate; they changed, and social capital changed with them.

> Because individuals—rather than households—are separately connected, the Internet and the cell phone have transformed communication from house-to-house to person-to-person. This creates a new basis for community that author Barry Wellman has called "*networked individualism.*" Rather than relying on a single community for social capital, individuals often must actively seek out a variety of appropriate people and resources for different situations (Rainee et al. 2006).

Activism depends on social networks and the effective use of social networking. However, before networks can develop, there must be a means of making and maintaining connections among those sharing the social change (or resistance to change) goal. With the loss of a localized, community basis—the town hall model—an alternative became necessary. Digital platforms have become that alternative—but to what effect?

Digital media has allowed millions of people to reconnect though common interests, to congregate together on a digital platform, and discuss news, issues and/or problems pertaining to the platform's designated topic. For example, an animal rights platform includes individuals who share strong opinions on animal rights, and who discuss current animal rights violations, possible solutions, and possibly also shared personal experiences related to animal rights. Thus, digital platforms allow individuals to form a slightly different form of social capital, because they no longer need to physically congregate in order to support a cause. Concerned netizens can discuss, blog, or comment on issues while in digital space. The current research analyzes eighty-five activism-oriented Internet platforms. A large majority of these (86%) included such discussion forums.

Creating social capital through digital media may ensure communication, but it does not guarantee results. It is important to remember that, although people communicate with more seemingly "like minded" individuals than ever before, it does not mean they are deeply committed to a cause they are discussing, or to the people they are discussing it with. Whether deeply committed or not, their participation through their digital presence does not necessarily change the status quo.

There is always the possibility that, even a lengthy cycle of exchanges may act to reinforce ideals without leading to material acts or contributing to concrete results. It is possible to have strong feelings about certain issues without moving beyond the phase of sharing perspectives and sentiments on certain topics. This produces nothing beyond reaffirming one's beliefs. Once that feeling is fulfilled, the individual moves on. Discussion is critical to networking. Digital platforms seem to offer serious potential for activism. So what are the problematic issues in activism through digital social networking? Three problematic dynamics are identified below: filter bubbles, relative visibility, and oversaturation.

Filtered Activism

The content of the Internet runs through several filters before being disseminated. Rather than information being distributed evenly about everything, it has adopted the same process of filtration as before; now, digital "gate-keepers" ultimately decide what information is being distributed. "There is this kind of shift in how information is flowing online, and its invisible, and if we don't pay attention to it, it could be a real problem" (Pariser 2011). Further:

> Filter bubbles are formed through generated signals in people's online queries. Filter bubbles refer to … your own personal-unique universe of information that you live in online … What's in your filter bubble depends on who you are, and it depends on what you do; but the thing is, you don't decide what gets in, and more importantly, you don't actually see what gets edited out (Pariser 2011).

For example, Facebook may edit out particular "news-feeds" from a person's friends list, because that Facebook member may have spent more time looking at other profiles, or responding more to news-feeds from other friends. As a person responds to messages and clicks on particular links from websites and member profiles, they contribute to massive data files; complex formulas are then used to create a unique profile that tailors future feeds and search responses to "fit" that profile. For example, there are fifty-seven signals Google has established in order to filter, personalize, and compartmentalize an individual's digital profile (Pariser 2011). Signals range from the type of computer one uses, to the location in which the person is accessing the information. Digital "gate-keepers" then create and store "filter bubbles" that become responsible for what information ultimately becomes disseminated to a particular individual. Thus, as people "surf" the net, click on links, view certain products, and read particular articles, their queries are recorded and become stored in large databases that are continuously updated and are instrumental in determining what information, links, and products will be available to them in future searches and queries (personalization). "The Internet is showing us what we think we want to see, but not necessarily what we need to see" (Pariser 2011). This process of filtration makes it very hard for people to watch or consume something that has not, in some sense, been tailored to them (Pariser 2011).

This is a significant problem for consciousness-raising, as anything not already part of a user's outlook is increasingly filtered away from view. Activists who are trying to recruit new supporters are increasingly likely to be trapped in the "preaching to the choir" mode. So the impact of informing content on activism platforms may be relatively ineffective despite the fact that all eighty-five such sites in the study sample attempt to engage in consciousness-raising through the dissemination of information. This particularly brings the utility of activism sites into question when recruiting potential cause supporters through information is their only function; this is the case for a small subset of the studied sites (14%). This customization also leads to activist forum discussions that serve to reinforce those already in agreement, but do virtually nothing to host persuasive discussions with those who may be unaware of the issues related to the cause touted. Almost one in five sites (18%) offer only information and discussion without other activism activity.

Currently, information outside someone's usual stream of filter-bubbled information is still *possible* to find and access. However, it takes *actual* effort. People can make intentionally explicit searches to find the exact material that might offer differing viewpoints, but not everyone is aware of how to go about doing so, or knowledgeable about filter bubbles, or motivated to seek out issues they may not yet know they would resonate with. Further, the culture of "likes" makes this type of effort increasingly unlikely. If one concedes that consciousness-raising may be the least effective use of SNS-based activism, what else presents a problem for activists online? Both relative visibility and oversaturation come into play.

Attraction and Distraction

Social networking sites may play a role in both attracting and distracting individuals. For example, Facebook has several hundreds to thousands of "causes" that flow fluidly across its site to attract their 2.2 billion users (Statistica 2018). Media outlets frequently divert attention away from significant social cause issues often disregarding pressing matters, and running with the most favorable or the most entertaining feed—those that will "trend" and attract concentrations of viewers. This will assure large numbers of clicks and continued attention. Mark Zuckerberg defended his deference to relatively unimportant infotainment by saying, "a squirrel dying in front of your house may be more relevant to your interests right now than people dying in Africa" (Pariser 2011). However, he did not address the fact that this is not a process of informed choice—users at least in part, have bids for their attention by what shows up in feeds and trending notifications. Visibility matters.

Relative Visibility

Filtering and the relative customizing of site visibility could possibly provide one benefit—reaching a pool of potential activists already sympathetic to a cause and therefore more likely to support it through donations. In fact, accepting donations was the only concrete activism function broadly available (81%) on the eighty-five sites studied. However, even so, the preeminence of sites vary. News and social cause feeds could easily be localized. However, when the number of clicks matter, feeds that have worldwide appeal will lead to more clicks and views than any local cause, no matter how pressing or engaging. Numbers translate into revenue for organizations hosting major digital platforms. Why does it matter if the focus is often on large-scale, generally "one-of" calls for aid? The outcome often over-supplies a single-crisis while drawing away needed support from ongoing issues.

For example, when the Gulf of Mexico was affected by an oil spill in 2010, people arrived from all over the country and donated large amounts of money to help. Meanwhile, the problem of toxic waste dumps in their local areas went unaddressed. Christens and Collura (2012) point out that concentrating too much on global issues displayed on social media may have a crippling effect on local causes.

Another form of visibility is critical for social change—the "feet on the ground" type of work required from supporters. This is where the digital platforms start to lose salience. Well under half of them (42%) provide access or contacts for volunteerism; just over one-quarter (28%) include petitioning connectivity; and only about one in five (19%) are at all tied to cause-supporting events—this includes even such minimal ties as announcing events or even where to get information on upcoming events.

Final Word

How does the advent of urban sprawl and ascendency of the Internet affect the social capital and material relationships with family, friends, neighbors, and co-workers? Consequently, how does this play out in activism and the hard work required for social change? Answers to these questions have long been debated:

> Those on one side of the debate extol the internet's ability to expand relationships—socially and geographically. Those on the other side of the debate fear that the internet will alienate people from their richer, more authentic relations (Rainee et al. 2006).

The fact that the geographic expansion of connectivity can be problematic for social action was discussed earlier. Perhaps, however, the social platform networking capabilities that dissipated social capital in communities can be turned around "to increase neighborhood social capital and the connectivity of local social networks" (Wellman and Hampton 1999).

In the early days of the Internet, Canada did an experiment with a residential network. They found social capital significantly increased in a neighborhood in which most residents had good Internet access and access to the residential network.

> Wired residents recognized three times as many of their neighbors, talked to those neighbors twice as often, visited them 50 percent more often, made four times as many local phone calls, and further boosted their local communication through the use of email (Wellman and Hampton 1999).

The findings were similarly positive for some basic elements of localized social activism. According to Wellman and Hampton (1999):

> Residents used their neighborhood networks to further build social capital by organizing local events (including house parties, block parties and barbeques) and to mobilize in dealing with community issues (including perceived housing deficiencies and problems).

Whatever promise this approach may have held was never realized. Perhaps networked individualism simply resonated too well with the individualism in long-standing American ethos. Perhaps too many other digital "attractions" compete for the attention of Internet users. Nonetheless, almost two decades later, residential networking has been slowly growing in the US. We have yet to see whether or not the earlier predictions of optimistic scholars come to pass:

> Residential networks become more prevalent, the Internet will increasingly be used to expand neighborhood social capital. Internet use may actually serve to reverse the decline of American social capital that has taken place over the last quarter-century (Wellman and Hampton 1999).

Whether or not the "wired neighborhood" becomes a viable platform for community connectivity, subsequent social capital, and local activism is questionable, given how quickly it faded from view more than twenty years ago.

Activism of any kind is information-reliant. Digital platforms were initially heralded by Americans as a means of democratizing access to information—especially once the smartphone closed the digital divide to a significant degree in the US (Lopez, Gonzalez-Barrera, and Patten 2013). This may be a short-lived victory, however. Presently, the Federal Communications Commission (FCC) has repealed Net Neutrality. This means people's access to information may be limited as time goes by. People would have to pay to enter certain websites, make searches online, and even access peer-reviewed journals for educational purposes. This of course would reintroduce polarized access to information, making it more difficult for middle or low-income families to afford Internet services. As a result, online memberships to certain entertainment and informational outlets may get more expensive, and information might not be as easily accessible. This author argues that such results show a situation of activism attrition on activism-oriented digital platforms. The greater the activism effort required, particularly any non-digital effort, the less effective the overall body of activism platforms seem to be in promoting or providing a venue for such efforts. This is another process that can be seen as cathartic, thus able to neutralize activism urges while seldom fueling sustained, committed, concrete action towards social change.

References

Christens, Brian D. and Jessica Collura. 2012. "Local Community Organizers and Activists Encountering Globalization: An Exploratory Study of their Expectations and Adaptations." *Journal of Social Issues* 68(3):592–611.

DeFronzo, James. 2011. *Revolutions and Revolutionary Movements*. Colorado: Westview Press.

Lopez, Mark Hugo, Ana Gonzalez-Barrera and Eileen Patten. 2013. "Closing the Digital Divide: Latinos and Technology Adoption." Pew Research Center: Hispanic Trends. March 7. Retrieved March 29, 2018 (http://www.pewhispanic.org/2013/03/07/vi-social-networking/).

Pariser, Eli. 2011. "Beware Online "filter bubbles." TedTalk. Retrieved May 22, 2015 (https://www.youtube.com/watch?v=B8ofWFx525s).

Rainee, Lee, John B. Horrigan, Barry Wellman, and Jeffrey Boase. 2006. "The Strength of Internet Ties." Pew Research Center. January 25. Retrieved May 21, 2018 (http://www.pewinternet.org/2006/01/25/the-strength-of-internet-ties/).

Rouse, Margaret. 2006. "Social Networking." Internet applications glossary. WhatIs.com.

Statistica. 2018. "Number of Active Monthly Facebook Users Worldwide As of 4th Quarter 2017." Retrieved April 25, 2018 (https://www.statista.com/statistics/264810/number-of-monthly-active-facebook-users-worldwide/).

Tonnies, Ferdinand, and Loomis, C.P. [1887] 2017. *Community and Society: Gemeinschaft und Geselleschaft*. Routledge.

Wellman, Barry, and Keith Hampton. 1999. "Living Networked in a Wired World." *Contemporary Sociology* 28(6):648–654.

Wirth, Louis. [1938] 2010. "Urbanism as a Way of Life." *The Chicago Journal of Sociology* 44(1):1–24.

Youmans, William L. and Jillian York C. 2012. "Social Media and Activist Toolkit: User Agreements Interests, and the Information Infrastructure of Modern Social, Corporate Interests, and the Information Infrastructure of Modern Social Movements." *Journal of Communication* 62(2):315–329.

Zilberstein, Karen. 2015. "Technology Relationships and Culture: Clinical and Theoretical Implications." *Clinical Social Work Journal* 43(2):151–158.

12

POWER AND MONEY

Explaining the Rise of Digital Media through Surveillance Capitalism

Wai Kit Choi

The Internet: Origins Matter

The rise of digital media has drastically changed how people around the world communicate and interact with one another; what could this mean for the future of human society? One way of answering this question is to suggest that digital technologies have both positive and negative effects on social interaction. To determine whether their overall impact is ultimately beneficial, we only need to calculate whether the "goods" outweigh the "bads." For example, from this commonsensical standpoint, the Internet, smartphones, social media platforms, big data, and the algorithms that are used to analyze the data collected can, on one hand, be used to establish closer ties between people and improve the efficiency of economic production and exchange. On the other hand, these different types of digital technologies and data are also thought to be contributing to mass surveillance and the infringement of their users' privacy. Based on the "compare and count" logic embodied in this perspective, one might refer to the general public's reception and conclude that digital technologies have more positives than negatives, since people continue to use and embrace them despite their threat to individual privacy.

This perspective is one way of thinking about the general implications of digital technologies, but it fails to capture the various complexities attendant to the issue. To illustrate, state policies and corporations' profit-making strategies play an important role in determining what type of technologies will be invented and how they will be used once invented. When examining the positive and negative impacts of digital technologies, this larger political-economic structure must be considered, and the commonsensical viewpoint overlooks this important factor while framing the deployment of technologies as simply a question of individual decision. In contrast to this approach, this chapter will give an analysis that accounts for the impacts of digital technologies by reference to the changing dynamics of political and economic development. Specifically, the concept of *surveillance capitalism* is central to this analysis. The first published work that explicates this concept thoroughly is the article by John Bellamy Foster and Robert W. McChesney (2014). Shoshana Zuboff uses the same concept in a speech that was published two months after Foster and McChesney's piece, and also in her subsequent publications on big data and Google Inc. (Zuboff 2015, 2016). These two uses of the concept overlap in some ways, but there are also differences. In both approaches, the term "surveillance capitalism" means a new phase of capitalist development where digital technologies have become ubiquitous (extremely common), and are used by corporations to collect personal data for the purpose of profit maximization. However, Foster and McChesney offer a historical narrative and underscore the interdependence between the state and corporations in their analysis, while Zuboff focuses primarily on the corporations and highlights the impacts of surveillance capitalism on autonomy and democracy. Despite their differences, both view surveillance capitalism as indicating a structural transformation in the political-economic system where the digitization of society has become all pervasive. Two causal processes can be extracted from their writings to explain what led to this structural transformation.

The first causal process is the emergence of a permanent war economy after World War II. Digital communication and surveillance capitalism would not have been possible had it not been for the Internet and other related technologies. Foster and McChesney link the inventions of these technologies to the Cold War—that is to the overt, geopolitical hostilities between Soviet, Eastern bloc countries and US-led Western powers from the end of World War II until 1990 (2014). They argue that the establishment of a militarized economy in the US that stimulated economic growth by incorporating civilian industrial production into war preparation, was what made these inventions possible. The second causal process is the growing dependence of the state and corporate sectors on digital technologies in their operations. The word *state* is to be distinguished from *government*. The former refers to the set of institutions that wields exclusive political authority over a territory, while the latter are the people who rotate in and out of the official positions within a state. Three institutions are particularly relevant: (a) the national security and military sector; (b) the corporate financial sector; and (c) the consumer research/marketing sector. As these institutions increasingly rely on surveillance and the collection of personal data as ways to maintain societal control or generate corporate profits, the digitization of our everyday lives becomes pervasive and hard to escape.

A Militarized Economy and the Internet

One of the most important factors that made the widespread application of digital technologies possible is the Internet. In Foster and McChesney's account, the way in which the Internet was invented was inextricably linked to the strategies that US state officials adopted to fight the Cold War and maintain capitalist economic expansion.

As World War II was drawing to a close, and the Allied victory appeared imminent, policy makers in Washington began to plan for future economic growth. Speaking to the special Congressional Committee on Postwar Economic Policy and Planning in 1944, the Assistant Secretary of State, Dean Acheson, warned that if the United Stated relapsed into an economic depression, such as the one that lasted from the 1920s to the 1930s, the consequences would be devastating. For Acheson, the economic downturn was not caused by the inability of firms to produce what people needed, but by the overproduction of goods—there was not sufficient demand for all that was produced. One strategy that successive postwar US governments adopted to prevent this problem, was the absorption of economic outputs by military spending (Foster and McChesney 2014). In 1946, while he was serving as the chief of staff of the Army, General Dwight D. Eisenhower issued a document titled *Memorandum for Directors and Chiefs of War Department General and Special Staff Divisions and Bureaus and the Commanding Generals of the Major Commands*. In this document, Eisenhower calls for the integration of the military with the civilian scientific and technological sectors. He notes that during armed conflicts, civilian industries are converted and become a crucial part of the war effort, but he adds that even during peacetime, civilian resources should remain "associated closely with the activities of the Army" (Foster and McChesney 2014:2). Eisenhower was suggesting that the partnership between the military and civilian industries formed during World War II should continue. This means that even when the US was not at war, the civilian economy could still find, in the armed forces, a demand for goods and services. We see in this document the beginning of an institutional arrangement that would later come to be called the *military-industrial complex*.

Eisenhower's focus was on strengthening the nation's war readiness by permanently integrating the civilian industries with the military; however, the idea that military spending could boost economic production, was explicitly expressed by other branches of the government. In 1952, President Harry Truman created the National Security Agency (NSA); just five years before that, the passage of the National Security Act also created the Council of Economic Advisers, the Central Intelligence Agency (CIA), and the National Security Council (NSC). From the beginning, the main responsibility of the NSC was to examine ways to defend US global domination in the face of growing Soviet influence. A set of strategies was laid out in a policy paper referred to as *National Security Council Report 68* (NSC-68), and one of its main recommendations was the strengthening of the military (National Security Council 1950). In the event of an armed conflict with the Soviet Union, the report pointed out, the US must be able to defend its allies in the Western hemisphere, protect communication lines, and conduct offensive operations (National Security Council 1950:54). The report then noted that:

> the ability to perform these tasks requires a build-up of military strength by the United States and its allies to a point at which the combined strength will be superior … to the forces that can be brought to bear by the Soviet Union and its satellites (National Security Council 1950:55).

Rearmament, as a strategy against the Soviet Union, required increased spending on the military and raised the question of whether there were adequate resources to implement the plan.

The answer from the report is that the:

> capability of the American economy to support a build-up of economic and military strength at home and to assist a build-up abroad is limited not, as in the case of the Soviet Union, so much by the ability to produce as by the decision on the proper allocation of resources to this and other purposes (National Security Council 1950:25).

For the architect of NSC-68, the challenge of bolstering the military was not the lack of resources, but the way resources were allocated. It is further pointed out that if the rearmament program was carried out, it "might not result in a real decrease in the standard of living, for the economic effects of the program might be to increase the gross national product by more than the amount being absorbed for additional military and foreign assistance purposes" (National Security Council 1950:58). This means that not only did the report reject the idea that increased spending on the armed forces was a drain on civilian resources, it saw, instead, a growing military budget as a stimulus to economic growth. The chairman of the Council of Economic Advisers, Leon Keyserling, was asked to weigh in on the NSC-68 recommendation; he also spoke of the economic benefits of a military build-up. Although civilian consumption would be affected, he noted that the impacts would not be severe while industrial outputs and employment would rise (Foster and McChesney 2014:5). An even more positive assessment is provided by Harvard University economist Sumner Slichter, who commented on the financial impacts of the Cold War, suggesting that "[m]ilitary spending increases the demand for goods, helps sustain a high level of employment, accelerates technological progress and helps the country to raise its standard of living" (Foster and McChesney 2014:6). For Slichter, an increased budget for the Cold War could bring economic benefits to the rest of society, like defense company contracts, and more employment opportunities in general.

Such comments from state officials illustrate that the Cold War with the Soviet Union, and the expectation that arms production could stimulate economic growth, are among the factors that helped permanently integrate armament build-up with the civilian economy. Early examples of the merger between these two sectors are President Eisenhower's selection of the Secretary of Defense, and the creation within the Defense Department the Advanced Research Projects Agency (ARPA), which was renamed as Defense Advanced Research Projects Agency (DARPA) in 1972 (DARPA nd).

Unlike all his predecessors, Eisenhower's appointee Charles Erwin Wilson, served a four-year term as the fifth US Secretary of Defense, without ever serving in the military; he was the President of General Motors prior to his appointment. Wilson's successor, Neil McElroy, also came from the business sector; before becoming Secretary of Defense, McElroy was the President of Procter & Gamble. These examples of corporate executives taking the helm at the Pentagon illustrate the crossover between the military and the civilian economic sector. This crossover was bolstered when the aforementioned ARPA was established in response to Soviet success in space. In 1957, the Soviet Union succeeded in launching two artificial satellites into space (*Sputnik 1* and *2*), and the US became concerned with falling behind in the technology race. Consequently, McElroy proposed the creation of ARPA as part of the armed forces within the Department of Defense, despite being a research agency that would consist mainly of scientists and experts from the civilian sector rather than military personnel (Foster and McChesney 2014:11). Private industries and universities would also be linked to ARPA as service or goods providers on a contract basis. Even the person whom McElroy picked to be the first director of ARPA, Roy Johnson, was from a business firm; he worked as a vice president at General Electric Company before joining the agency.

Following approvals by Eisenhower and Congress, ARPA began operation in 1958, and the research that it has since undertaken included the development of spy satellites (the Corona Reconnaissance Satellite program), the world's first global satellite positioning system (the Transit program), stealth technology for combat aircrafts, the Predator drone, and numerous other projects.

From ARPANET to Internet

The work that is perhaps the most relevant to the digital background discussion here is the Advanced Research Projects Agency Network (ARPANET), which is the precursor of today's Internet (DARPA website). A number of different factors converge to make the ARPANET possible.

First, the unit within ARPA responsible for research on ARPANET was the Information Processing Techniques Office (IPTO), and its first director was J. C. R. Licklider. However, this office was initially called Command and Control Research, and was created when the Director of Defense Research and Engineering transferred a surplus computer from the Air Force to ARPA, assigning them the task of developing command and control studies (Barber 1975:49). In military parlance, *command and control* refers to the communication and interaction between the different elements embedded within the same organizational structure; these elements include "all institutional structures, physical systems, and human resources utilized in the execution of military strategy" (Wood 2006:121). During the Cold War, the US government was concerned with the possibility of a nuclear attack from the Soviet Union. As a result, the Institute for Defense Analyses determined that the military command and control system must link to the following functions:

1 Maintain an up-to-date accounting of the status of forces and nuclear weapons.
2 On the defensive side, secure as early warning as possible of an enemy attack, assess it, and pass that warning to the National Command Authorities and to the strategic forces.
3 Communicate the order to launch strategic forces and maintain contact with them after launch.
4 Ascertain the effectiveness of strike forces and the restrike capability of those forces.
5 Maintain the capability to carry out these functions during and after a nuclear attack on the United States (Wainstein et al. 1975:XI).

Through the creation of this new office, ARPA would conduct research on improving the interaction between the different components of the national security command and control system.

Second, Licklider was made the director of the Command and Control Research office in 1962. Prior to his appointment, Licklider worked on an air defense system called Semi-Automatic Ground Environment (SAGE) for the Air Force, and conducted research on human-computer interaction. One of his papers, titled "Man-Computer Symbiosis," was highly influential and is seen as a one of the key documents that inaugurated the shift from treating computers as merely computational devices to viewing them as multifunctional machines that can also serve as communication tools. Perhaps more importantly, Licklider was interested in studying communication between computers; he proposed the concept of *galactic network*, which is a "globally interconnected set of computers through which everyone could quickly access data and programs from any site" (Leiner et al. 2009:23). With Licklider in charge, the military's concern with developing a command and control system as part of a nuclear retaliatory capability converged with research on computer-to-computer communication. Changing the name of the office to IPTO indicates this communicative turn, while ARPANET can be seen as the attempt to initiate his idea of galactic network.

Third, while Licklider's innovative ideas served as the conceptual foundation for ARPANET, Paul Baran and colleagues proposed the key technological breakthrough (packet switching) for network communication. In 1959, Baran joined the Rand Corporation, an organization originally created by the Air Force to conduct research on national defense issues. Computer simulations at Rand showed that the national telephone system would be rendered inoperable following a nuclear strike. Baran was responsible for designing a survivable communication system. His solution was to embed data transmission within a *distributed network*, which is a network "without any hierarchical structure; thus, there is no single point of vulnerability to bring down much of the network" (Baran 2002:41). During transmission, a message is then divided into packets that take different routes and are reassembled at a destination. In 1967, Lawrence Roberts became the leader of the IPTO's ARPANET project and recruited Baran after learning about his work (O'Neill 1995:78; Baran 2002:48). It is important to note that Baran's research on packet switching, similar to the creation of IPTO (the command and control research office) in ARPA, was prompted by military war preparation efforts. Explaining the history of the ARPANET, the former Director of ARPA, Stephen Lukasik, notes:

> Writing from the viewpoint of the person who signed most of the checks for Arpanet's development, in this article, I detail the rationale for investing US Department of Defense resources for research and development of the first operational packet-switched network. The goal was to exploit new computer technologies to meet the needs of military command and control against nuclear threats, achieve survivable control of US nuclear forces, and improve military tactical and management decision making (2011:4).

The ubiquitous expansion of digital media in our society today is inextricably linked to the invention of the Internet, the origin of which is traced back to the ARPANET. Foster and McChesney show how the hostilities between the Soviet Union and the US, as well as the American policy makers' concern with averting the capitalist crisis of overproduction, led to the establishment of a permanent war economy. A key feature of such an economy, as illustrated by the creation of ARPA within the Department of Defense, is the fusion of the military with the civilian technological/industrial sector. The ARPANET is a product of this historically specific institutional arrangement. The connection between war preparation and the invention of ARPANET is also underscored by Lukasik, who sees ARPA's commitment of financial resources to this research project as an attempt by the US military to computerize, or digitize, its struggle against the Soviet Union.

One way of explaining the prevalence of technology is to attribute to it an inherent usefulness that allows it to satisfy the need of the general population; once a critical mass of people learn about what this technology can do and adopt it, the technology becomes widespread. However, this explanation cannot be so easily applied to digital media. As the history of ARPANET shows, the digitization of everyday life began with the digitization of war. Bringing convenience to the general population's daily life was *not* the goal—the goal was bringing destruction to the enemy.

Explaining Digitization through the Push Factors

This argument that the military-industrial complex and war preparation are key factors contributing to the spread of digital technologies may have critics. Some may object that a distinction can be made between the initial invention and later applications. In other words, while the motivation behind inventing a technology is not about fulfilling a need or solving a problem common to the general population, that invention later takes on a life of its own and is applied and reappropriated in different social spheres and environments. In this way, one sees the unintended consequences of the invention. For these critics, even if digital technologies were not initially invented to meet the needs of the masses, they were later extended to serve that purpose. For example, Google, Facebook, Apple, Amazon, Uber, Skype and many other corporations constituting the digital economy today all offer products that purportedly make their users' lives more convenient. As the number of interested users grow, and their everyday activities become deeply integrated with these products, society becomes increasingly digitized.

There is undoubtedly some truth to this outlook; however, it inflates the influence that individual users have over the course of societal digitization. This leads back to the importance of digital technology's military-industrial roots. Critics who ignore this factor end up overlooking how the state and corporations are able to use their immense political and financial power to push for a digitized socio-economic order—one where individual involvement is not an option but a requirement. For example, once a smartphone is acquired, a purchase that is increasingly hard to avoid for work or social reasons, the users become easily ensnarled in a host of pre-installed mobile apps. In fact, they even need to create a Google or Apple account otherwise the software on their phones cannot be updated. The prevalence of digital technologies, therefore, cannot always refer to their success in satisfying some supposedly pre-existing needs of the users, the *pull factors*. Rather, we need to also examine the *push factors*—the dynamics that lead to a digitized socio-economic order that has become nearly impossible to avoid. This means that, even in later applications, the larger political-economic institutional context still matters since the state and corporations play a crucial role in promoting digitization. The concept of surveillance capitalism is also important in this connection.

Expanding State and Corporate Power Through Surveillance

Compiling information on the activities of potential *subversives* through surveillance has long been a strategy used to maintain political control. *Subversives* (those whose activities express opposition to the state) and potential subversives can now be tracked and profiled as never before. What makes digital technologies attractive is that they can make the collection of personal data far more comprehensive and systematic. From the standpoint of the state, this is one reason for expanding digitization and integrating more and more people into the web of digital data management. For corporations, personal data are no less important. In the digital economy, the services provided by many of the tech companies such as Google, YouTube, Yahoo!, Facebook … etc. are free; in exchange, they monitor the users' activities, and derive profit from the data they collect. It is important for these corporations to expand digitization since the increase in users will give them more personal data, which leads to higher profit growth.

Foster and McChesney examine some of the ways in which the US state and corporations conduct surveillance and discuss how such practices have become digitized. The state surveillance programs that they refer to include the FBI's Counterintelligence Program (COINTELPRO), the Army's Continental United States intelligence program (CONUS), and the NSA's various mass surveillance programs exposed by Edward Snowden in 2013. The targets of COINTELPRO were individuals whom the FBI saw as a danger to the government—primarily activists and, in some cases, journalists. The methods that the FBI employed to monitor their activities or dismantle their organizations included "a wide array of surveillance and illegal activities (break-ins, forgeries, agent-provocateur actions, wrongful imprisonment, and violence)" (Foster and McChesney 2014:15). In a 1976 report prepared by a US Senate Select Committee, the famous "Church Committee," these FBI tactics are referred to as "indisputably degrading to a free society" (United States Senate Select Committee to Study Governmental Operations with Respect to Intelligence Activities 1976:10). The Army's CONUS intelligence operations were also a domestic surveillance program that targeted civilians engaging in political activities (New York Times 1971). Military units were deployed to quell mass demonstrations in cities during the 1960s, and the Army maintained that its creation of the spying program arose from the need to predict future civil unrest (United States Senate Subcommittee on Constitutional Rights Committee on the Judiciary 1973:4). The agencies that were responsible for gathering data were the US Army Intelligence Command (USAINTC) and Continental Army (CONARC), while the Counterintelligence Analysis Branch (CIAB) was tasked with analyzing and predicting future uprisings. Digital technologies in their early development were used in these operations. Computers were set up to store and analyze data in order to plot "trends and discern incipient civil disorders" (United States Senate Subcommittee on Constitutional Rights Committee on the Judiciary 1973:62). The computerized *Incident Data Files* tracked various "incidents" that were captured by using ninety-six categories such as "strike," "march," and "convention" (United States Senate Subcommittee on Constitutional Rights Committee on the Judiciary 1973:62). An index number was also assigned to each of the persons and groups contained in the file. Through an index number, an intelligence officer can identify all of the organizations that a particular person belonged to, or all the members in a particular organization. Another set of computerized files was the *Biographical Data File* that had information on 4,078 persons including "priests, lawyers, singers, comedians, politicians, and relatives of federal officials (including one member of the family of a US Senator), state legislators, former federal officials, leaders of the Urban League, the Black Panthers, [and] the Black Muslims" (United States Senate Subcommittee on Constitutional Rights Committee on the Judiciary 1973:62–63). Following the investigation of the CONUS intelligence program by Congress, the Army was told to destroy the files. Instead the files were transferred to the NSA via the ARPANET; this is an illustration of the use of a "proto-Internet system" in state surveillance (Foster and McChesney 2014:14).

As shown by Edward Snowden's disclosure of the NSA's spying programs in 2013, the digitization of state surveillance has become much more extensive today than during the 1970s, and this transformation underscores two characteristics of surveillance capitalism in particular. First, previous spying operations covered thousands of people who were considered potential subversives. Digital surveillance today follows a different logic and millions of people are now being monitored, most of whom show no sign of involvement in political or criminal activities. For example, the NSA was collecting the metadata of millions of Verzion phone users. These data could indicate which calls were made from a given number, the duration of each call, and the location of the callers and recipients (Greenwald 2013). An enormous amount of data is needed for big data analysis, a method that allows intelligence agencies to identify patterns of behavior and the correlations among them. As a former CIA officer explained, the importance of a piece of information might not become obvious until it is combined with data to be obtained in the future. This means that intelligence agencies have collected all kinds of information from as many people as possible in order to hold onto everything they have for as long as they can (Sledge 2013). Second, Snowden's files on the NSA, an intelligence unit within the Department of Defense, shows the continuation of the relationship between the military and the corporate sector that enabled the development of the ARPANET during the Cold War. Examples of military-industrial collaboration include:

- An NSA program codenamed, Prism. The spy agency paid Microsoft, Google, Yahoo, Facebook, Skype and other companies for data on their users which were made available in different formats—emails, voice/video chats, file transfers, online social networking details, *etc.* These companies initially denied their involvement, then claimed that their participation was due to a compulsory court order. The NSA claimed that this program cost $20 million a year and claimed that their analysts could make use of "collection directly from the servers" of US service providers (MacAskill 2013; The Guardian 2013).
- A program called Fairview that involved AT&T. Based on the documents which date from 2003 to 2013, it is reported that AT&T gave NSA access to billions of emails that were traveling through its domestic networks, and that in 2011, they also started providing over 1 billion cellphone records made in the US to the spy agency on a daily basis. With the help of AT&T, the NSA was also able to conduct surveillance on the United Nations headquarters, and the telecom company was praised for its "extreme willingness to help" and was described as "highly collaborative" (Angwin et al. 2015).
- An NSA PowerPoint presentation titled "Corporate Partner Access" that emphasized the importance of having a good relationship with tech companies. The NSA also obtained data from international communication; to do that, they needed access to fiber-optic cables, and this required the cooperation of the telecom companies that owned the networks of underwater cables (MacAskill and Rushe 2013).

This brief post-World War II US history of state surveillance practices shows an unmistakable trend toward more comprehensive digitization for the purpose of data collection. In fact, the relationship between digital technologies and surveillance can be described as a feedback loop—the development of these technologies create new possibilities of surveillance, while the growth in surveillance, in turn, leads to a greater need for digitization. Private companies also incorporate surveillance in their day-to-day operations, and also help expand the use of digital technologies in society.

For Foster and McChesney, the business sector's turn to surveillance is a result of the *financialization* of the US economy. This phenomenon, where finance and financiers have increasingly achieved dominance over the rest of the economic system, can be traced back to the Reagan era when the government deregulated the financial sector, and investors began turning to finance (rather than manufacturing goods) for greater profitable returns. To capture the funds flowing into the financial market, various speculative investment instruments were invented; simultaneously, the insurance industry and the personal finance sector—which covered credit cards, store cards, auto loans, student loans, home mortgages, and many other areas—also expanded. In this context, surveillance becomes important for the financial industry, given its concern with analysis-based predictions and risk management. Consequently, the data on personal incomes, spending habits, credit, and other types of personal information are now being collected, stored and analyzed. This growing need for mass personal data, which stems not only from the financial but also the consumer research/marketing sector, gave rise to data brokers—firms that specialize in the collection of personal information. One of the biggest companies in this industry, Acxiom, has a total market value of over $2 billion (Yahoo Finance 2018). Foster and McChesney note that Acxiom has:

> 23,000 computer servers processing in excess of 50 trillion data transactions annually. It keeps on average some 1,500 data points on more than 200 million Americans, on the form of "digital dossiers" on each individual, attaching to each person a thirteen-digit code that allows them to be followed wherever they go, combining online and offline data on individuals (2014:19).

Like state intelligence agencies, Acxiom also gathers big data on people, and it is just one of the many companies that are conducting digital surveillance.

Big Data as a Commodity

In Zuboff's analysis, surveillance capitalism is also conceptualized as a new phase in capitalist development where corporations make profits through the collection of mass personal data. But unlike Foster and McChesney, Zuboff does not discuss the relationship between the US military and corporations. She attributes a much more central role in inaugurating and disseminating this new logic of profit accumulation to the digital technology sector, Google Inc. in particular. "Google" she writes "is to surveillance capitalism what Ford and General Motors were to mass-production and managerial capitalism a century ago: discoverer, inventor, pioneer, role model, lead practitioner, and diffusion hub" (Zuboff 2016). The company began to assume this pivotal role when it was confronted with the need to increase profit in 2001 following the crash of the dot-com bubble, and created "the most successful scheme for making money on the Internet that the world had ever seen" (Levy 2011:81).

Google's method of generating revenue was a new advertising model that had two characteristics. First, Google's customers are not the people who use its search engine, email service or web browser. Rather, they are the business firms that purchase advertisements placed within Google's search result pages and other web-based products that are widely used all over the world. Second, the advertising model that Google adopts is often referred to as *targeted advertising*. Unlike TV commercials that promote their products to all viewers indiscriminately, Google ensures the effectiveness of its method by matching advertisements on its search result pages to people who are more likely to be interested in the products. As Vaidhyanathan notes, many companies, especially small firms

> don't need to scream at millions of people that they should be buying some brand of weak beer. They need to attract the attention of potential consumers who have expressed interest in, say, Bavaria. For this reason, Google needs to understand how patterns of searches indicate behaviors. If Google can customize the placement of ads, giving a user results listing only local shoe stores or only Bavarian lager, then it can generate more clicks per advertisement (Vaidhyanathan 2012:27).

The customized advertising placement that Vaidhyanathan refers to here is key in enabling Google to win over its corporate clients. On the basis of web users' IP addresses or other identifying information (such as their Google accounts), Google can identify patterns of online behavior after saving and analyzing the histories of their queries on its search engine. But in addition to search history, Google also stores and analyzes other types of data on web users' online behavior. For example, many digital publishers and advertisers hire an ad-serving firm called DoubleClick to manage the advertisements on their websites and analyze their effectiveness; a recent study that examines tracking on the top 1 million websites shows that DoubleClick is used on about half of these sites (Englehardt and Narayanan 2016:1395).

A Trail of Cookies

One of DoubleClick's methods of managing advertisements is the use of tracking cookies. When people visit a website that uses a DoubleClick service called Adsense, text files, *cookies*, are placed on their browsers. These cookies record the web users' browsing activities, and they stay on their browsers even after they leave the sites. DoubleClick continues to track these users as they go to other websites that are part of the Adsense network; the cookies on these visitors' browsers are activated, the ads that they click on, the videos they watch, and the articles they read at all these different sites are recorded (Geary 2012). The data on a web user's activities at different web sites will be analyzed, and a profile indicating the user's behavioral pattern or general interest—for example, whether the user is a sport fan or an art enthusiast—can be created to help determine advertisement placement. What makes DoubleClick particularly relevant to our discussion here is the fact that Google purchased the company for $3.1 billion in 2008. Through Google Play Store, Gmail, and the company's search engine, Google already has the names of its numerous account users, their phone numbers, and their search history. With the additional data from DoubleClick, it has become even more difficult for web users to keep any secrets from Google (Angwin 2016).

Under this surveillance-based business model, Zuboff notes that web users are "neither buyers nor sellers nor products" (Zuboff 2016). Rather, the intimate details of their daily lives are now the raw materials upon which data analysis is performed, and they are then transformed into prediction profiles that Google uses to generate profit. As more companies in Silicon Valley and other business sectors adopt the same profit-making strategy, this surveillance-based business model becomes generalized throughout the economy. Zuboff's examples of consumer products that collect data on their users include sport apparel, "smart vodka bottles," and "Internet-enabled rectal thermometers" (Zuboff 2016).

Manipulated by Our Profiles

In response to Zuboff, one can argue that people who volunteer information about themselves on the social media and are not concerned about the loss of privacy would not find surveillance capitalism a threat. However, the problem with digital surveillance is not merely the collection of data on our daily activities, hence our loss of privacy, but also the ways corporations use these data. As Zuboff points out, companies are using data about us to modify our behavior for profit. Targeted advertising represents one type of behavioral modification since it uses the personal data collected on web users to sway their shopping decisions. Auto insurance companies can track their customers' driving habits through telematics devices installed in cars; based on the data collected, which may include the customers' location, driving speed, distance driven, cell phone usage while driving, etc., these insurance companies modify their customers' behavior through "punishments (real-time rate hikes, financial penalties, curfews, engine lock-downs) or rewards (rate discounts, coupons, gold stars to redeem for future benefits)" (Zuboff 2016). A scientist at a Silicon Valley firm that develops educational applications, expresses clearly in an interview with Zuboff the relationship between data collection and behavioral modification:

> [The] goal of everything we do is to change people's actual behavior at scale. When people use our app, we can capture their behaviors, identify good and bad behaviors, and develop ways to reward the good and punish the bad. We can test how actionable our cues are for them and how profitable for us (Zuboff 2016).

As it is explained here, profit is generated from using data to change people's behavior. When firms in different business sectors all adopt a similar business model, entrapment seems to prevail in all spheres of social existence. For Zuboff (2015, 2016), the challenge that surveillance capitalism poses is not just the erosion of privacy, but also the disappearance of self-determination and autonomy, principles that are the basis of a democratic society.

Conclusion

In discussion about the popularity of digital media or the pros and cons of its impacts on society, individual preference is often invoked as a key consideration. For example, it can be argued that digital media is popular simply because it brings convenience to people's lives and, as more and more people find it useful and incorporate it into their daily activities, our society becomes thoroughly digitized. Furthermore, the general public continues to embrace digital media despite the challenges that it poses, including the invasion of privacy. One can argue that, for most people, usage rates indicate that the benefits of a digitized society outweigh its problems.

However, this chapter shows that this approach to explaining digital media by reference to individual preference is inadequate. Rather than explaining the ubiquity of digital media solely in terms of the needs or preferences of individual users—pull factors—the concept of surveillance capitalism underscores the importance of push factors. The state and corporate push for thorough societal digitization can be seen as a strategy to maintain power and generate profit, hence rendering the use of digital media unavoidable for the general population. State and corporate strategies expand digitization throughout society in two ways. First, the US state or, more specifically, the US military pioneered and played a key role in the invention of the Internet, without which the diffusion of digital media would not have been possible. The Cold War prompted the military's involvement based on the ideas that the US state was in an arms race with the Soviet Union, and that the nation's war readiness had to be maintained in the event of an actual conflict. To address these challenges, policy makers integrated the military and the civilian economic sector. This collaboration, also referred to as the military-industrial complex, gave the former access to resources and talents outside the government while giving the latter an outlet for its products. An example of this collaboration was the establishment of ARPA (or DARPA) within the Department of Defense.

Researchers from corporations, think tanks, and universities were hired on a contract basis to work on specific projects. One of these projects was the ARPANET, predecessor of the Internet, developed to ensure that military communication systems would remain functional even after a nuclear attack by the Soviet Union. The rise of digital media was enabled by the Internet, and as the ARPA example demonstrates, its invention originates in US military plans to use it in the power struggle against the Soviet Union.

Second, the US state does not expand digitization only through the development of the Internet. For both the state and corporations, the collection and analysis of personal information has become a central part of their operations, and both have increasingly used digital technologies to facilitate these efforts. In the case of the state, the maintenance of social and political control requires identifying potential challengers, or subversives, accomplished, in part, through gathering mass personal data on the general population. As illustrated by the NSA files that Snowden leaked to the media, the NSA's mass surveillance programs are only made possible by data-management systems that are thoroughly digitized. Surveillance is equally important to corporations. Under the business model pioneered by Google, corporate profit is generated by the collection of detailed information on people's daily activities, and is achieved by incorporating as many people as possible into the digital universe.

All too often, the popularity of a technology is seen as a reflection of its success in satisfying the needs of the users. Focusing on digital media, the explanation offered here emphasizes, instead, roles played by the state and corporations in creating an everyday order where nearly all spheres of our existence—social, political and economic—are enacted through digital technologies. In so far as participation in such an everyday world is not a choice but a necessity, people will inevitably adopt these technologies regardless of what they actually prefer.

References

Angwin, Julia. 2016. "Google has Quietly Dropped Ban on Personally Identifiable Web Tracking." *ProPublica*, October 21. Retrieved May 15, 2018 (https://www.propublica.org/article/google-has-quietly-dropped-ban-on-personally-identifiable-web-tracking).

Angwin, Julia, Charlie Savage, Jeff Larson, Henrik Moltke, Laura Poitras, and James Risen. 2015. "AT&T Helped US Spy on Internet on a Vast Scale." *The New York Times*, August 15.

Baran, Paul. 2002. "The Beginnings of Packet Switching: Some Underlying Concepts." *IEEE Communications Magazine* 40(7):42–48.

Barber, Richard J. 1975. "Advanced Research Projects Agency Report, 1958–1974." (Declassified Report) Associated Inc.

DARPA. nd. "History and Timeline." Defense Advanced Research Projects Agency. Retrieved March 12, 2018 (https://www.darpa.mil/about-us/darpa-history-and-timeline).

Englehardt, Steven, and Arvind Narayanan. 2016. "Online Tracking: A 1-Million-Site Measurement and Analysis." Proceedings of the 2016 ACM SIGSAC Conference on Computer and Communications Security: 1388–1401.

Foster, John Bellamy, and Robert W. McChesney. 2014. "Surveillance Capitalism: Monopoly-Finance Capital, the Military-Industrial Complex, and the Digital Age." *Monthly Review* 66(3):1–31.

Geary, Joanna. 2012. "Doubleclick (Google): What Is It and What Does It Do?" *The Guardian*, April 23. Retrieved September 19, 2018 (https://www.theguardian.com/technology/2012/apr/23/doubleclick-tracking-trackers-cookies-web-monitoring).

Greenwald, Glenn. 2013. "NSA Collecting Phone Records of Millions of Verizon Customers Daily." *The Guardian*, June 6. Retrieved September 19, 2018 (https://www.theguardian.com/world/2013/jun/06/nsa-phone-records-verizon-court-order).

The Guardian. 2013. "NSA Prism Program Slides." November 1. Retrieved September 19, 2018 (https://www.theguardian.com/world/interactive/2013/nov/01/prism-slides-nsa-document).

Leiner, Barry M., Vinton G. Cerf, David D. Clark, Robert E. Kahn, Leonard Kleinrock, Daniel C. Lynch, Jon Postel, Larry G. Roberts, and Stephen Wolff. 2009. "A Brief History of the Internet." *ACM SIGCOMM Computer Communication Review* 39(5):22–31.

Levy, Steven. 2011. *In the Plex: How Google Thinks, Works, and Shapes Our Lives.* New York: Simon and Schuster.

Lukasik, Stephen. 2011. "Why the ARPANET Was Built." *IEEE Annals of the History of Computing* 33(3):4–21.

MacAskill, Ewen. 2013. "NSA Paid Millions to Cover Prism Compliance Costs for Tech Companies." *The Guardian*, August 23. Retrieved September 19, 2018 (https://www.theguardian.com/world/2013/aug/23/nsa-prism-costs-tech-companies-paid).

MacAskill, Ewen, and Dominic Rushe. 2013. "Snowden Document Reveals Key Role of Companies in NSA Data Collection." *The Guardian*, November 1. Retrieved September 19, 2018 (https://www.theguardian.com/world/2013/nov/01/nsa-data-collection-tech-firms).

National Security Council. 1950. *NSC 68: United States Objectives and Programs for National Security*. (Declassified Report).

New York Times. 1971. "A Shift Reported in Surveillance." February 24. Retrieved September 19, 2018 (https://www.nytimes.com/1971/02/25/archives/a-shift-reported-in-surveillance-senators-hear-justice-unit-takes.html).

O'Neill, Judy E. 1995. "The Role of ARPA in the Development of the ARPANET, 1961–1972." *IEEE Annals of the History of Computing* 17(4):76–81.

Sledge, Matt. 2013. "CIA's Gus Hunt On Big Data: We 'Try to Collect Everything and Hang On To It Forever'." *Huffington Post*, March 20. Retrieved September 19, 2018 (https://www.huffingtonpost.com/2013/03/20/cia-gus-hunt-big-data_n_2917842.html).

United States Senate Select Committee to Study Governmental Operations with Respect to Intelligence Activities. 1976. *Intelligence Activities and the Rights of Americans: The Church Report*. April 26. Retrieved September 19, 2018 (http://www.thirdworldtraveler.com/FBI/Church_Committee_Report.html).

United States Senate Subcommittee on Constitutional Rights Committee on the Judiciary. 1973. *Military Surveillance of Civilian Politics*. Washington DC: US Government Printing Office.

Vaidhyanathan, Siva. 2012. *The Googlization of Everything:(And Why We Should Worry)*. Berkeley, CA: University of California Press.

Wainstein, L., C.D. Cremeans, J.K. Moriarty, and J. Ponturo. 1975. *The Evolution of US Strategic Command and Control and Warning, 1945–1972* (Declassified Report). Institute for Defense Analyses.

Wood, Jason D. 2006. "Survival of the Fittest: The Evolution of US Military Command and Control Structures During and After the Cold War." *Comparative Strategy* 25(2):121–131.

Yahoo Finance. 2018. "Acxiom Corporation." Retrieved July 18, 2018 (https://finance.yahoo.com/quote/ACXM/key-statistics?p=ACXM).

Zuboff, Shoshana. 2015. "Big Other: Surveillance Capitalism and the Prospects of an Information Civilization." *Journal of Information Technology* 30(1):75–89.

Zuboff, Shoshana. 2016. "The Secrets of Surveillance Capitalism." *Frankfurter Allgemeine*. May 3. Retrieved May 20, 2018 (http://www.faz.net/aktuell/feuilleton/debatten/the-digital-debate/shoshana-zuboff-secrets-of-surveillance-capitalism-14103616.html).

13

KNOWING YOU BETTER THAN YOU KNOW YOURSELF

Manufacturing Perceptions

Michael Nitzani

The Questions

What happens when "big data" are used to anticipate, customize, and substantially reshape and control our online experiences? How do they know so much about us? What are the implications for our overall social agency? How might we "fight back?" This chapter discusses the information needed to answer these questions.

How They Know You

The digital world is inhabited by people of all ages, but the millennial generation came of age in a world of video games, computers, and applications that have created new dimensions to socially interact with one another. This is the first generation of digital natives—one that has constantly been connected to the Internet—for Millennials, the Internet and networked society is no longer a technological innovation, but a way of life.

Technology companies like Facebook, Google, Amazon, Reddit, Twitter, Snap, Apple, Microsoft, and Netflix IAC (Tinder, OkCupid, Plenty of Fish, and Match.com), have all contributed to dramatically changing the way individuals interact in the social world. They have created large networks and tools that people rely on to communicate with one another, find opportunities, discover new ideas, and promote easier access to consumer goods and services. These companies have become embedded in the process of connecting people with other people, places, and things. As a result, these companies have helped create pervasive, dynamic networks that have reshaped the socio-structural pathways through which people start, build, and maintain relationships with one another.

Each of these companies uses sophisticated data operations to encourage user engagement with their networks. This data is recorded and, when combined across networks, is known as big data. With big data, there is no need for hypotheses and guesswork, just statistical correlations (Schönberger and Cukier 2014). Companies have discovered that in mass datasets, there are data points that have predictive qualities of future behavior. These large-scale patterns and trends can be analyzed as a whole, and then targeted to an individual in order to influence a person's future habits. To illustrate the intrinsic value of this data for businesses, the Target Corporation can potentially tell that a woman is pregnant before even she knows, by identifying certain purchases she makes in comparison to those of other pregnant women with similar purchasing patterns (Duhigg 2012). This type of big data use is called predictive analytics.

There is a rush for data that can identify individuals, not only across all their social media and SNSs, but across digital devices. This type of data is collected from the growing set of Wi-Fi connected objects. For example, the Oral-B 7000 has a tooth-brushing app that allows its user to track brushing usage, "earn" trophies, and share "achievements" on Facebook. Data companies are able to work across the variety of such Wi-Fi connected devices, like digital assistants, vacuums, thermostats, cars, refrigerators, and more, to extract personal information, and then generate new information about our personalities. This is referred to as data mining. Most users are unaware of how much data their Wi-Fi connected things communicate and transmit; they also know little about how these items serve as latent surveillance tools that can reveal their most intimate behaviors and routines. This is how Google Analytics Solutions can promise business clients "website, app, digital, and offline data to gain customer insights." Their website includes this statement:

> Google Analytics gives you the analytics tools you need to analyze data from all touch points in one place, for a deeper understanding of the customer experience. You can then share the insights that matter with the whole organization (Google Analytics Solutions nd).

Many of the agreements users "accept" in order to access an online application, allow companies to obtain user data. There is no serious mechanism for individuals to "opt out." There is little concern over being complicit when opting in. Consent to these legal agreements does not guarantee the safe keeping of personal information. Rather, it allows access to the summary of your individual data, and "consent" often denies users the right to sue. Scripts, cookies, and web beacons, track our digital usage across the Internet. Applications track our data as we move through both the digital and the physical world. All of this data is stored inexpensively and indefinitely, so companies can create digital composites of individual user actions and choices. They access this data to create individualized offerings that appeal to us, and keep us engaged with their services.

If Target can recognize a woman's pregnancy from a few purchases, imagine what companies can do with the thousands of data points they amass. From our "likes," and other digital behaviors, models of our gender, ethnicity, political views, intelligence, happiness levels, age, and sexual orientation can all be inferred pretty accurately (Kosinski, Stillwell and Graepel 2013). With newer advances in artificial intelligence, companies can examine information across unstructured data sets like status updates, forum postings, reviews, and other meta data, and apply "machine learning" algorithms to identify patterns. This is combined with individual user data. On one hand, this data has created a sociologically fertile way to understand human experience and connections. On the other hand, this very type of understanding provides opportunities to take advantage of human psychological vulnerability in ways the world has never seen.

Attention-Seeking Economy

Computing technology provides platforms for creating immense social networks that become embedded in social interaction. Digital networks enable users to discover, explore, and seek out a range of content and possibility. They keep people easily connected with one another. Companies create intuitive ways to organize a wealth of information conveniently that keeps people reliant on their services. In many ways, these companies have created insular bubbles for how people access and process information.

Technology companies compete with each other for user attention by using aggressive algorithms (formulas) to serve up what people are interested in, along with what companies think they are interested in, based on what they have visited before, their "likes," and what advertisers are willing to pay to have them see. Napster creator and early Facebook investor, Shawn Parker explained the thinking behind the design process for the Facebook platform:

> The thought process was about how do we consume as much of your time and conscious attention as possible? … That means we need to give you a dopamine hit every once in a while because someone liked or commented on a photo or a post, or whatever. And that's going to get you to contribute more content … It's a social validation feedback loop. It's exactly the kind of thing that a hacker like myself would come up with because you're exploiting vulnerability in human psychology. (Allen 2017)

Distributed in random intervals, this "digital dopamine" is highly addictive and a part of an attention-seeking economy where user attention is auctioned off to the highest bidder.

On social media platforms, online status is represented in friend and follower ratios, and distributed in "likes," "loves," and "favorites." These are subtle affirmations that people use to give feedback to one another. For many people, in exchange for their posts regarding their personal information, inner thoughts, and experiences, social validation is ample enough payment. For people who have large followings, their visibility allows them to be "branded" and contacted with opportunities. It can also make their friend lists a target. One familiar example of this is when an organization tries to increase traffic to its site by putting messages on an individual's Facebook pages to announce that a "friend" has visited or "liked" a particular site.

An adaptation to these large social networks is that people may employ digital protections akin to those described for distinguishing private physical presence as "backstage" (Goffman 1959). This is done by using accounts that purposefully limit one's connections and filter one's interactions. On the other hand, what might generally be considered to primarily belong to the physical realm of "backstage" behavior has become digital front stage behavior for others, like camming. Camming literally means to live stream candid footage of casual, personal, or even sexual behavior over a significant period of time while talking to or chatting with various viewers. This digital behavior that can be monetized with little to no overhead on platforms like YouTube, Twitch, and Chaturbate. Some people utilize the platforms for attention, and others don't wish to bring attention to their alternate identities; data-mining algorithms operate without being able to act on this distinction—they do not "understand" that not everybody wants to be "connected." In fact, to illustrate the need for privacy, sex workers operating in digital space have been encouraged to "friend" their clients so they have greater control over client privacy (Hill 2017a).

Social media platforms are also designed around engagement loops. In appeals to broader audiences online, people and entities will use platforms to respond with their own inputs to establish their own voice and perspective on current events. To build user engagement, many websites will allow unified logins (giving users simultaneous access to all their SNS from a single point), which facilitates analytics from companies like Twitter, Facebook, and Google, because it effectively extends the site's analytical data reach. This allows companies to receive instant registration data on who is visiting their sites. For example, PornHub data mainstreams and reveals "hidden" sexual desires in their programming to ensure maximized engagement (Illing 2017). Given the nature of their site, however, PornHub implies significant knowledge—while simultaneously downplaying how much they actually know—about individual users:

> Thanks to the anonymized data provided by Google Analytics, Pornhub's statisticians are able to build an accurate picture of the

> demographic makeup of our visitors including their gender, age and even interests (PornHub 2018).

Another example is the autoplay algorithm that predicts what "you might also like."

The YouTube algorithm is notorious for autoplay content that gets progressively more extreme as a result of highly sophisticated deep neural networks that track engagement analytics and autoplay recommended content (Lewis 2018). YouTube is known for paying YouTubers who generate high enough viewer traffic to attract advertisers. These YouTubers also earn through direct viewer sponsorships, or commercial endorsements.

With people and entities competing for these view-based revenues, the autoplay recommendation engine has created numerous controversies in regard to free speech. Autoplay can bring viewers to sites they were not aware of—and each click matters. YouTubers who earn money with viewer numbers do not want their speech curtailed, especially if it is the more outrageous content can garner more views. In this sense, "free speech" is speech that appears "for free," but might actually generate revenue because it can become attached to advertising. One YouTuber, Logan Paul, was suspended by the platform for a series of behaviors that included filming a suicide victim, tasering a dead rat, and a tweet (later deleted) encouraging his followers to do a potentially lethal "tide pod challenge" by swallowing laundry detergent packets (Vincent 2018).

Recently, disturbing videos targeting children with representations of popular "kid's show" characters engaged in nonsensical and emotionally disturbing behaviors are also being posted (Bridle 2017). It is not clear if these videos are made by humans, artificial intelligence, piggybacking off of page views, or something more nefarious. The rampant extent of these videos is revealed in an official Google blog post, where CEO of YouTube, Susan Wojcicki, touts a machine-learning AI that serves to remove hundreds of thousands of videos with "extremist" content. However, for inappropriate videos that make it past the state-of-the-art machine-learning algorithms, YouTube promised that 10,000 human moderators would manually vet content (2017).

In exchange for the conveniences of hosting online calendars, fostering search queries, getting directions, and connecting people with what they are interested in, users unwittingly generate immense data sets that reveal intimate details about them. Promotions, including screen messages, encourage people to download applications (apps) because they offer convenience; however, these apps grant revealing data to developers. From apps, guided pathways frame polite requests for sensitive, personally identifying information like your contact lists, email lists, and any other information that has conceivable business value. With this information, analysis can infer numerous things about an individual's life (Dewey 2016). Often though, people freely give this up, without conscious thought. Companies can use this data and correlate it with things like geo-spatial correlations to spur further connections—forming shadow profiles with information that is collected about people who do not even have accounts on a platform (Hill 2017b). They also sell highly intimate personal information (Pierson 2018). When the Strava fitness app released a global heat map, mapping out the routes where people have gone, a student tweeted how he was easily able to identify sensitive military bases (Hern 2018). Twitter's engagement algorithms sent the tweet to reporters. This news instantly became viral and was picked up by other media organizations. The ensuing fallout encouraged Strava to make the opt-out process much simpler (Locklear 2018).

Yet, for people concerned with their user data, these companies often employ heavy persuasion tactics to automatically pull people into invasive privacy agreements and settings. When end users signal that they are no longer willing to be financial contributors to these websites by using an advertising blocker, or wish to assert their "privacy preferences," their decisions are undermined with "are you sure?" messages and pop-ups that may even include pictures of a person's friends. When someone is not engaged, they will send notifications for attention, use confusing pathways to obscure opt-out information, and bury important terms and settings in legalese and doublespeak.

A Post-Modern Narrative on Digital Democracy

Digital text takes on a life of its own. Whether text is targeted advertising, or is based on backlinks, the correlations in which it objectively appears can easily be manipulated, misinterpreted, and misunderstood.

Click and Comment Value

To earn enough views to appeal to advertising, there is value in clicks. There are people who write for paid clicks, using search engine optimized language (SEO). This is a language in and of itself, as it is intended to appeal to common search terms rather than substance. A different technique to generate clicks uses clickbait, which are headlines written with language that either withholds information or greatly exaggerates its claims. This was popular on social media sites until Facebook changed their algorithms as a result of anti-clickbait sentiments (El-Arini and Tang 2014). However, these tactics are ever-present in today's digital media environments. Even venerable newspaper organizations will use SEO tactics to gain readers (Holiday 2017). Journalists are encouraged to examine website analytics to put their articles up at opportune times for user engagement (Christin 2017). This means of appealing to audiences encourages journalists to write on topics they know their audience will read, generating content that people will engage with more, at the expense of important, but less compelling stories (Holiday 2017).

For many years, written text superseded spoken text as a means of authority. When people read a stranger's opinions on digital platforms, it can be difficult to discern if they are truthful, are acting on behalf of brands or business, or are even automated bots. Internet companies, however, look only at engagement metrics—it does not matter where the ideas come from, whom they represent, how credible they are, or what intent they serve. A few concentrated voices can hijack engagement algorithms and affect the way that information appears in digital environments. Sensitivity to user rates and user preferences means that popular social networking sites enable users to manipulate what will be seen—what is visible to others—resulting in messages that can be artificially amplified. On Instagram, "engagement pods" will comment and like each other's posts so that way they appear in each other's feeds and boost exposure (Thompson 2017). On Reddit, brigades will band together to upvote and downvote particular comments, and on Twitter, users will band together to launch certain tweets to create a "meticulously manufactured virality," earning some users thousands of dollars a month (Reinstein 2018). Like a poker player bluffing with a bad hand, these presentations can influence other people's perceptions with real consequences.

With automated bots, messages can appear to be much more popular than they really are, challenging the role that tech companies play in public discourse and the dissemination of information (DiResta 2018). In a tweet, Twitter's CEO, Jack Dorsey, articulates the difficulty this new environment has caused the platform.

> We have witnessed abuse, harassment, troll armies, manipulation through bots and human-coordination, misinformation campaigns, and increasingly divisive echo chambers. We aren't proud of how people have taken advantage of our service, or our inability to address it fast enough (Dorsey 2018).

There is an incentive to manufacture reviews which appear legitimate in order to reach higher rankings or appeal to potential clients. To illustrate the faultiness of these automated review algorithms, a British prankster detailed how he tricked Tripadvisor.com by writing numerous overwhelming positive reviews for his backyard shed, ultimately becoming the top ranked restaurant in London (Butler 2017). These reviews are subjective. Numerous websites have issues with their ranking systems. Positive reviews can be paid for through astroturfing or sock puppeting (Hancock 2012).

Astroturfing keeps the identity of a sponsor hidden to make it appear like the message comes from an average user, rather than an organization with an agenda (commercial, political, religious, etc.). Similarly, sock puppeting is using a deceptive digital identity, generally to talk about oneself, while pretending to be someone else—like a coffee house manager writing a review of his own place, while pretending to be a customer. Bad reviews are less likely to exist, great reviews are average, and on sharing platforms, good reviews are inflated (Fowler 2017; Ziegele and Weber 2014). On the flip side, negative reviews, whether they are true or not, can be debilitating to companies, or to people, as a negative perception can affect their place in a search algorithm.

Manipulation, Filters, and Memes

These algorithms are influential. Facebook caused substantial controversy in 2014 when researchers highlighted that the site could manipulate people's emotions with their friend feed via emotional contagion where the sentiments that a person sees in their feed affects their outward sentiments (Kramer, Guillory, and Hancock 2014). Since then, Facebook has routinely run experiments in their news feeds, using responses to opposites of similar content, colors, trends, ideas, and language analysis, then they determine from a person's "likes" who they get information from, what appeals to them, and their perspectives on issues so that Facebook can create a highly-personalized news feed to connect them with people that they think they are interested in, and hyper-target this individual to advertisers by specific micro category.

The flip side of data-responsive personalizing and customizing is limiting and homogenizing what is experienced. Search engine records and big data algorithms combine to influence which sites come up, in what order, in subsequent searches. The more searches, the more customized the search results—until one person's search results differ significantly from the results of another person making the same search. This filter bubble effect impacts what things that are visible online to whom.

During the US election of 2016, these individualized feeds would create filter bubbles that served up engaging and polarizing content stemming from likes and personality profiles. These would constitute a feed with content designed to appeal to user interests. These feeds would frame media choices based on socio-psychological profiles and data-driven associations. These filter bubbles would encourage people to connect with those with similar viewpoints and perspectives for further engagement.

Because of unified logins across the Internet, and the way people could organize into groups, Facebook served as a quasi-public space for social interactions. Facebook users would engage in vigorous arguments with people who had opposing viewpoints. Caught in the middle of a debate about persuading voter opinion, Facebook wanted to appear impartial. In response to criticism that they received from conservative groups (that their trending stories section had human editors), the company fired their editors and allowed their automated algorithms more leeway in choosing trending topics during the run-up to the election (Thompson and Vogelstein 2018).

When the context of language is removed and replaced, it changes the meaning of the original message. With digital tools, it is easy to reframe ideas into easily understandable bite-sized chunks and create new narratives. When combined with imagery, in the form of memes, these messages can be persuasive, despite being factually incorrect or misleadingly out of context. Mobilized websites and applications will often omit details to reformat text to fit on the smaller screen. When text is reformatted across digital devices, users have different perceptions on what is seen as important. Nevertheless, when these ideas scroll past, the mind still perceives them, and misperceptions filter into subconscious thoughts (Kahneman 2015). Whether it is valid or not, psychologists argue that merely processing this information with minimal awareness initially retains it as "truth" (Gilbert 1991). Due to the decentralized nature of online text, there are few consequences for digital deception. Digitized democracy on the Internet, means that anybody can say anything authoritatively, even if it is not true. (Vosoughi, Roy and Aral 2018).

In the case of fast-moving events, automated engagement algorithms will push false stories to the forefront as these situations create a high demand for related information. Ideas spread easily with digital text because the form and content can easily be separated and syndicated (Wesch 2007). Salacious headlines, pictures that appeal to user biases, and money from third party advertising networks like Google Adsense, help grab the views. The advertising/attention economy relies on the fact that people are more willing to create, react, and share the things that provoke them, and they believe will be provocative enough to bring attention to them and their take on the topic. Data scientist Josh Schwartz confirmed that "articles that get a lot of tweets don't necessarily get read very deeply. Articles that get read deeply aren't necessarily generating a lot of tweets" (Manjoo 2013). The more extreme the position, the more likely it is to go viral. The good money potential makes creating fake news a serious business (Subramanian 2017).

Recent history provided an arena where predictive analytics and fake news vied for viewer attention. Cambridge Analytica (a political consulting firm) used machine-learning algorithms to conduct experiments on the effect of advertising specific messages to individual users based on psychological profiles generated through big data. Their goal was to persuade them to vote for their clients in the 2016 US election (Anderson and Horvath, 2017). The company's vice president of Global Media, Molly Schweickert detailed how Cambridge Analytica developed a carefully coordinated digital advertising campaign (using readily available technologies) to hyper-target individual voters with "strategic messages" (2017). Concurrently, however, the Russian-based Internet Research Agency used an army of trolls, including automated bots, to comment angrily on forums, create memes, and interact with each other. Their goal was also to shape people's opinions in the 2016 US election. While we don't know the overall effectiveness of these campaigns, it is apparent that the reach of psychological manipulation is unprecedented as political entities attempt to, and no doubt do, influence a significant number of people.

The Quandary

Internet users have been led to engage in digital space or through digital platforms where lucrative opportunities exist for people who capitalize off "likes" and views. People and machines communicating in digital space environments can dramatically affect life in the physical world. Messages can appear and disappear by a tweak of an algorithm that shapes what is visible to individual users. Thousands of data points about each of us exist, and will continue to exist, allowed by opaque privacy policies and bought and sold by digital data brokers. This data will continue to be hacked and mismanaged.

With the massive online opportunities for people to express themselves, practice deception, manipulate others, de-contextualize events, and conduct communication "experiments," understanding the true nature of specific digital information is increasingly problematic. Given the sheer amount of content, this requires a new digital literacy to combat pressures for engagement, political biases, and manipulated narratives.

Implications and Advice

The Revolution Happened

With the current environment of daily-platform use environment, and the inclusion of alternative facts, fake news, and truthiness, American society has officially entered something that resembles a post truth, post-modern, post-structural era. The attention economy encourages viral misinformation. Our lives are embedded in multiple digital environments, subject to the serious big data formulas and filter bubbles that influence the information presented to us. It is increasingly necessary to evaluate and reconcile our actions in the interconnected digital and physical world.

Law, data use, and privacy need to be reconsidered in view of automated artificial intelligence, and big data algorithms, that increasingly influence and govern our lives. Academics and scholars worry that we are becoming dependent on artificial intelligence without understanding the processes used to make decisions (Kuang 2017). Artificial intelligence is getting better at picking up the nuances in datasets, but it is still a form of abstracted empiricism that does not entirely understand people—especially the often self-contradictory nature of human behavior. Despite filtering and predictive analytics, there is still ingenuity to human behavior, and a randomness that even the most powerful super computers cannot comprehend.

However, responsible adults need to consider how these new technologies—that can literally manufacture perceptions—might affect the social world. To put things in perspective, the current invasive data collection and advertising model will ultimately shift just as previous models have shifted in the past. In some ways this is already clear through examples such as the impending European Union General Data Protection Regulations. As more people become aware of their data, predictive analytics business models may be challenged. Companies may be forced to employ easy opt-out strategies or risk financial penalty. Annual reports from companies like Alphabet (parent of Google), Facebook (Instagram, WhatsApp), Twitter, and Snap have revealed that they may have oversold the effectiveness of the hyper-targeted advertising model.

For Your Consideration …

Facial recognition software, artificial intelligence, and digital means to augment concrete realities are now happening. Current technology can already create: holograms of Tupac and others (Tsukayama 2012), computer voices that are indistinguishable from human voices (Gershgorn 2017), software that manipulates visual faces in real time (Plaugic 2016), humanoid robots (Redmon 2017), deepfake videos using software that substitutes one face for another in a visually undetectable way (Kharpal 2018), smart glasses, and open source chat bots incorporating a real individual's personality (Pardes 2018) with eventual abilities to form their own language (Metz 2017).

Mostly Disengage

Deconstruct the images and ideas presented to you online and take stock: Is it a "this or that" quiz to test your preferences? Are you being experimented on? What do your digital breadcrumbs reveal? How many times you are prompted for action when you do engage with these companies?

Consider how your notifications affect your concentration. Are they all that important? Some may be. Take some time to limit the notifications that are not. Evaluate whether you are in a 24/7 relationship with a digital device and periodically be offline and fully engaged in physical space. Digital devices cannot interrupt and engage with you if there are periods where they cannot find you. When you engage with these companies, map out exactly what you want to do—resist falling down too many rabbit holes.

Do More Thinking for Yourself

Life is complex and complicated, yet our decisions are becoming simplified through filter bubbles and SEOs. Our thought processes should not be. We need to actively seek information from competing factions to place our own ideas in a larger narrative.

Digital devices are embedded into our realities. However, you can take some time away from your devices. Appreciate your own thoughts and anxieties. It allows you to see bigger picture ideas. It allows you to put pieces in the puzzle together. You are less vulnerable to influence when you can complete your own thoughts without distraction. You are most likely to be influenced or distracted when your thoughts are scattered. Less automation means the more you to do things manually. This is not necessarily a bad thing on occasion. You are less likely to react to prompts given to you from networked influences, when you are more in tune with your directly inhabited physical space environment.

Reclaiming Agency

People adapt to their environments. Human beings are not automatons. It is possible to reject terms of agreement. Take stock. What does your data reveal about you? Consider the amount of connected devices you have, the behaviors that they monitor, and the way that you are prompted for engagement. Are you OK with the data that you grant these companies? You are the one who grants them permission to use your data. You can employ "adblock" and even take the means to control your host files on your devices to limit what they can find out about you from your usage. You can set up filters for the emails that they send you, dummy names for the names that you register with, and pay for things with cash more often. You can jump through their hoops to regain more privacy.

Keep big data in mind. You do not need to be necessarily honest, and in most cases you are not obligated by law to use your real information. Big data can still access your information, it just dirties up their data a bit, which might throw off their computer models of behavior. Consider the notion that you do not have to "like" the things that you like. You can mess with the machine to decouple the digital space personality you have exhibited. Think about what sites and activities they suggest to keep your attention—knowing that automated algorithms are processing numerous things and reconciling what they know about you and your digital connections—these suggestions indicate what they think they know about you. You can start choosing in ways to confuse or contradict those "knowns."

See the World Beyond

The rise of misinformation requires that we actively try to understand, not only what we see in digital space, but the mechanisms that brought it into our line of sight. We need to look for opinions outside of what filtered and mediated interaction encourages.

As more entities in digital environments attempt to influence your thoughts, do what you can to regain the actual context of social phenomenon. Read books. Talk with people. Observe the social cues in physical environments. You have the power to change your own habits. You have the freedom to get lost and rediscover the world around you. To connect with people face-to-face—to trust other human beings. What the computers cannot predict are your real, visceral, weird, random, human experiences with others. They might not be able to detect your skepticism or sarcasm (Bamman and Smith 2015). To better avoid manipulation, you need to step back and analyze behaviors outside of the contexts in which they are delivered to you. You may even have to challenge your own thoughts. Change your mind. See what happens when you do. When you take some time away from digital devices, you are walking out of the metaphorical casino full of cognitive traps designed to keep you in. You are reclaiming your own experience. When you get back to the connected world, keep in mind for perspective, that the words that exist, and ideas that are propagated, are data-driven and largely exist to appeal solely to you. You are going to need to hold onto this awareness as the artificial intelligence you engage with levels up.

References

Allen, Mike. 2017. "Sean Parker Unloads on Facebook: God Only Knows What it's Doing to Our Children's Brains." *Axios*November 9. Retrieved March 15, 2018 (https://www.axios.com/sean-parker-unloads-on-facebook-god-only-knows-what-its-doing-to-our-childrens-brains-1513306792-f855e7 b4-4e99-4d60-8d51-2775559c2671.html).

Anderson, Berit and Brett Horvath. 2017. "The Rise of the Weaponized AI Propaganda Machine." *Scout*February 9. Retrieved January 31, 2018 (https://scout.ai/story/the-rise-of-the-weaponized-ai-propaganda-machine).

Bamman, David and Noah A. Smith. 2015. "Contextualized Sarcasm Detection on Twitter." Pp. 574–577 in Proceedings of the Ninth International AAAI Conference on Web and Social Media. Austin: Association for the Advancement of Artificial Intelligence.

Bridle, James. 2017. "Something Is Wrong on the Internet." Medium November 6. Retrieved April 4, 2018 (https://medium.com/@jamesbridle/something-is-wrong-on-the-internet-c39c471271d2).

Butler, Oobah. 2017. "I Made My Shed the Top-Rated Restaurant on TripAdvisor." *Vice*December 6. Retrieved April 4, 2018 (https://www.vice.com/en_us/article/434gqw/i-made-my-shed-the-top-rated-restaurant-on-tripadvisor).

Christin, Angèle. 2017. "Algorithms in Practice: Comparing Web Journalism and Criminal Justice." *Big Data and Society* 4(8):1–14. Retrieved April 4, 2018 (http://journals.sagepub.com/doi/pdf/10.1177/2053951717718855).

Dewey, Caitlin. 2016. "98 Personal Data Points That Facebook Uses to Target Ads to You." *The Washington Post*August 19. Retrieved April 4, 2018 (https://www.washingtonpost.com/news/the-intersect/wp/2016/08/19/98-personal-data-points-that-facebook-uses-to-target-ads-to-you/?utm_term=.b4206642d153).

DiResta, Renee. 2018. "The Information aWar is On. Are We Ready for It?" *Wired*August 8. Retrieved March 19, 2019 (https://www.wired.com/story/misinformation-disinformation-propaganda-war/).

Dorsey, Jack. 2018. "Twitter." *@Jack.* Retrieved March 1, 2018 (https://twitter.com/jack/status/969234279321419776).

Duhigg, Charles. 2012. "How Companies Learn Your Secrets." *The New York Times*February 16. Retrieved February 20, 2018 (http://www.nytimes.com/2012/02/19/magazine/shopping-habits.html).

El-Arini, Khalid and Joyce Tang. 2014. "Click-Baiting." Facebook Newsroom August 25. Retrieved April 4, 2018 (https://newsroom.fb.com/news/2014/08/news-feed-fyi-click-baiting/).

Fowler, Geoffrey A. 2017. "When 4.3 Stars Is Average: The Internet's Grade-Inflation Problem." *The Wall Street Journal*April 5. Retrieved January 31, 2018 (https://www.wsj.com/articles/when-4-3-stars-is-average-the-internets-grade-inflation-problem-1491414200).

Gershgorn, Dave. 2017. "Google's Voice-Generating AI is Now Indistinguishable from Humans." *Quartz*December 26. Retrieved February 20, 2018 (https://qz.com/1165775/googles-voice-generating-ai-is-now-indistinguishable-from-humans/).

Gilbert, Daniel T. 1991. "How Mental Systems Believe." *American Psychologist* 46 (2):107–119.

Goffman, Erving. 1959. *The Presentation of Self in Everyday Life.* New York: Double Day.

Google Analytics Solutions. nd. "Improve the Customer Experience with Digital Analytics." Retrieved April 30, 2018 (https://www.google.com/analytics/analytics/#?modal_active=none).

Hancock, Jeff. 2012. "The Future of Lying." *Ted Ideas Worth Spreading*September. Retrieved January 31, 2018 (https://www.ted.com/talks/jeff_hancock_3_types_of_digital_lies/transcript).

Hern, Alex. 2018. "Fitness Tracking App Strava Gives Away Location of Secret US Army Bases." *The Guardian*January 28. Retrieved April 4, 2018 (https://www.theguardian.com/world/2018/jan/28/fitness-tracking-app-gives-away-location-of-secret-us-army-bases).

Hill, Kashmir. 2017a. "How Facebook Outs Sex Workers." *Gizmodo*October 11. Retrieved April 4, 2018 (https://gizmodo.com/how-facebook-outs-sex-workers-1818861596).

Hill, Kashmir. 2017b. "How Facebook Figures Out Everyone You've Ever Met." *Gizmodo*November 7. Retrieved April 4, 2018 (https://gizmodo.com/how-facebook-figures-out-everyone-youve-ever-met-1819822691).

Holiday, Ryan. 2017. *Trust Me, I'm Lying: Confessions of a Media Manipulator.* New York: Portfolio/Penguin.

Illing, Sean. 2017. "Proof that Americans Are Lying about their Sexual Desires." *Vox*June 27. Retrieved February 20, 2018 (https://www.vox.com/conversations/2017/6/27/15873072/google-porn-addiction-america-everybody-lies).

Kahneman, Daniel. 2015. *Thinking, Fast and Slow.* New York: Farrar, Straus and Giroux.

Kharpal, Arjun. 2018. "Reddit, Pornhub Ban Videos that Use AI to Superimpose a Person's Face Over an X-Rated Actor." *CNBC*February 8. Retrieved February 20, 2018 (https://www.cnbc.com/2018/02/08/Reddit-pornhub-ban-deepfake-porn-videos.html).

Kramer, Adam D.I., Jamie E. Guillory, and Jeffrey T. Hancock. 2014. "Experimental Evidence of Massive-Scale Emotional Contagion Through Social Networks." *Proceedings of the National Academy of Sciences* 111(24):8788–8790.

Kuang, Cliff. 2017. "Can AI Be Taught to Explain Itself?" *The New York Times*November 21. Retrieved April 4, 2018 (https://www.nytimes.com/2017/11/21/magazine/can-ai-be-taught-to-explain-itself.html).

Kosinski, Michal, David Stillwell, and Thore Graepel. 2013. "Private Traits and Attributes are Predictable from Digital Records of Human Behavior." *Proceedings of the National Academy of Sciences* 110(15):5802–5805.

Lewis, Paul. 2018. "'Fiction Is Outperforming Reality: How YouTube's Algorithm Distorts Truth." *The Guardian*February 2. Retrieved April 4, 2018 (https://www.theguardian.com/technology/2018/feb/02/how-youtubes-algorithm-distorts-truth).

Locklear, Mallory. 2018. "Strava Simplified How to Opt out of Its Heat Map Data Collection." *Engadget*March 1. Retrieved April 4, 2018 (https://www.engadget.com/2018/03/01/strava-simplified-opt-out-heat-map/).

Manjoo, Farhad. 2013. "You Won't Finish This Article." *Slate Magazine*June 6. Retrieved April 4, 2018 (http://www.slate.com/articles/technology/technology/2013/06/how_people_read_online_why_you_won_t_finish_this_article.html).

Metz, Cade. 2017. "It Begins: Bots Are Learning to Chat in Their Own Language." *Wired*March 16. Retrieved April 4, 2018 (https://www.wired.com/2017/03/openai-builds-bots-learn-speak-language/).

Pardes, Arielle. 2018. "The Emotional Chatbots Are Here to Probe Our Feelings." *Wired*January 31. Retrieved April 4, 2018 (https://www.wired.com/story/replika-open-source/).

Pierson, David. 2018. "Gay Dating App Grindr Changes Its Policy of Sharing Users' HIV Status with Outside Vendors." *Los Angeles Times*April 2. Retrieved April 4, 2018 (http://www.latimes.com/business/technology/la-fi-tn-grindr-hiv-data-20180402-story.html).

Plaugic, Lizzie. 2016. "Watch a Man Manipulate George Bush's Face in Real Time." *The Verge*March 21. Retrieved February 20, 2018 (https://www.theverge.com/2016/3/21/11275462/facial-transfer-donald-trump-george-bush-video).

PornHub. 2018. "2017 Year in Review." Pornhub Insights January 9. Retrieved September 19, 2018 (https://www.pornhub.com/insights/2017-year-in-review).

Redmon, Joseph. 2017. "How Computers Learn to Recognize Objects Instantly." *Ted Ideas Worth Spreading*April. Retrieved September 19, 2018 (https://www.ted.com/talks/joseph_redmon_how_a_computer_learns_to_recognize_objects_instantly).

Reinstein, Julia. 2018. "'Tweetdecking' Is Taking Over Twitter. Here's Everything You Need To Know." *BuzzFeed*January 12. Retrieved April 4, 2018 (https://

www.buzzfeed.com/juliareinstein/exclusive-networks-of-teens-are-making-thousands-of-dollars?utm_term=.yoQo3ny61#.kh7krjWBm).

Schönberger, Victor and Kenneth Cukier. 2014. *Big Data: A Revolution That Will Transform How We Live, Work, and Think.* Boston, MA: Mariner Books, Houghton Mifflin Harcourt.

Schweickert, Molly. 2017. "How Digital Advertising Worked for the US 2016 Presidential Campaign." D3con May 12. Retrieved April 4, 2018 (https://www.youtube.com/watch?v=bB2BJjMNXpA).

Subramanian, Samanth. 2017. "The Macedonian Teens Who Mastered Fake News." *Wired*February 15. Retrieved April 4, 2018 (https://www.wired.com/2017/02/veles-macedonia-fake-news/).

Thompson, Rachel. 2017. "'Selfie Decks' Are the Hot New Trend You're Not Invited To." *Mashable*May 9. Retrieved April 4, 2018 (https://mashable.com/2017/05/09/selfie-decks-twitter-groups/#6bCiV74HA5qf).

Thompson, Nicholas and Fred Vogelstein. 2018. "Inside the Two Years that Shook Facebook and the World." *Wired*February 12. Retrieved April 4, 2018 (https://www.wired.com/story/inside-facebook-mark-zuckerberg-2-years-of-hell/).

Tsukayama, Hayley. 2012. "How the Tupac 'Hologram' Works." *The Washington Post*April 18. Retrieved February 20, 2018 (https://www.washingtonpost.com/business/technology/how-the-tupac-hologram-works/2012/04/18/gIQA1ZVyQT_story.html).

Vincent, James. 2018. "YouTube Suspends Logan Paul's Ad Revenue, Blames 'Recent Pattern of Behavior'." *The Verge*February 9. Retrieved April 4, 2018 (https://www.theverge.com/2018/2/9/16994364/logan-paul-youtube-suspends-advertising).

Vosoughi, Soroush, Deb Roy and Sinan Aral. 2018. "The Spread of True and False News Online." *Science* 359(6380):1146–1151. Retrieved September 19, 2018 (https://doi.org/10.1126/science.aap9559).

Wesch, Michael. 2007. "The Machine is Us/ing Us (Final Version)." YouTube March 8. Retrieved September 19, 2018 (https://www.youtube.com/watch?v=NLlGopyXT_g).

Wojcicki, Susan. 2017. "Expanding Our Work against Abuse of Our Platform." YouTube: Official Blog December 4. Retrieved April 8, 2018 (https://youtube.googleblog.com/2017/12/expanding-our-work-against-abuse-of-our.html).

Ziegele, Marc and Mathias Weber. 2014. "Example, Please! Comparing the Effects of Single Customer Reviews and Aggregate Review Scores on Online Shoppers Product Evaluations." *Journal of Consumer Behavior* 14(2):103–114.

Part IV

CAPPING IT OFF

Preface

This final part of the book is a capstone. Chapter 14, "Why We Care," reframes the book in view of the authors' intentions. Rather than a rehashing of the previous thirteen chapters, overarching considerations are addressed. Additional insights are offered from the quoted material of the abstinence journal entries. The commentary addresses reader awareness, and suggests avenues for introspection. The primary purpose is to provide an integrated reflection on the overall implications raised, and the issues that should be kept in mind.

Central to this chapter is a somewhat conflict theory perspective asking readers to consider the chapters in view of the ongoing struggle among individual identity development, socializing forces, and social control in the context of co-presence in digital space and physical space. More specifically—readers are asked to consider digital life manifestations of autonomy and empowerment; tensions among digital avenues of socialization, social agency, and social justice as they interrelate to powerful digitally-supported institutional interests, and systems of social control. Ultimately, this short chapter attempts to communicate one of the most salient, overall "take-aways" from this book. It is also self-reflective, asking readers to consider whether or not the authors succeeded in presenting the need for mindful netizenship.

Chapter 15 is not a simple collection of methodology sections from the various authors. This chapter provides a methodological reflection that builds on, but also bridges the research projects supporting this book. As such, it provides avenues of discussion concerning the overall merit of several different methodologies applied by the authors, in broad terms. Each author included at least some form of original, empirically based work exploring digital life and human experience; however, this is not where the chapter begins.

Chapter 15 starts by defining and discussing the primary outlooks informing the research process. Constructionism and phenomenology are unpacked, and then explored, in terms of how they frame the set of original research activities from which the book draws. This includes considering of the importance of the subjective perspectives of study participants, the role of life as they experience it, the need to be mindful of the stable or changing factors contextualizing the phenomenon, and the relationship between findings, and support for politicized viewpoints, social change, or policy.

Second, overall process is addressed. From a process perspective, the importance of rapid social change is noted as a factor shaping research. The usefulness of context is noted in this regard, as is the need for grounded theory. Instructive commentary on grounded theory is included. Such commentary incorporates the definition and relative use of deductive and inductive reasoning processes, and the iterative aspects of the research endeavor. The approach is also treated as one in which contextualized and situational factors are given their due.

Third, specific methods are addressed. The explication is organized by the primary method involved, rather than by chapter. The goal is not to critique the specifics of research undertaken in *individual articles*, but to consider the value of the varied methodological approaches using authors' research as *concrete examples*. Aside from one survey-based project, all other methodologies are qualitative—although the minor blurring of such boundaries is discussed in view of standardized data collection using qualitative material that is then consistently coded and yields numerical comparisons. Most authors have included emergent research practices, as explained and illustrated in this chapter. The various specific methods considered for relative strengths and challenges are: phenomenological social-history, digital autoethnography, time-interval diaries, four types of content analysis (audiovisual, hypertext, iconic, and written), and self-administered surveys.

14

WHY WE CARE

Netizenship and Informed Choice

Cristina Bodinger-deUriarte

Interfacing Ourselves Through Mindful Co-Presence

All established societies involve an ongoing struggle among three interconnected arenas of human concern: (1) individual identity, autonomy and empowerment; (2) socialization, social agency, and social justice; and (3) powerful institutional interests, and systems of social control. Historically, these have played out within physical space communities, domestic institutions, national agencies, and international organizations. This has changed with the advent and rapid expansion of digital-space platforms and near total Internet access in the US. The "world wide web," social media, virtual subculture communities, instantaneous messaging, and digital commerce, provide a wealth of possibilities to the individual, but within a problematic set of circumstances:

- Powerful social-psychological dynamics may lead to individual over-engagement, ineffective multi-tasking, or physical presence neglect, unbalancing other important aspects of the co-present life.
- The unique visible/not visible context of digital space lessens the sense of embarrassment and self-monitoring common in face-to-face interactions, resulting in boundary violations, over-sharing, and other toxic interactions that lead people to seek advice about their digitally-mediated relationships with family, intimates, friends, and "unfriends."
- Digital netiquette and virtual group dynamics tend to encourage norms of immediate response and escalating connectivity demands, that may eventually become oppressive or unbalance co-presence.
- Unfettered capitalism and sophisticated manipulative elements built into apps and other interactive and tracking features—especially in social media platforms—may over-favor commercial interests while undermining individual access to balanced information.

- True civic representation and informed social agency are further eroded through increasingly customized and filtered browser responses and AI manipulations.
- Conjoined capitalistic and governmental interests in data-mining, along with the capability of wireless devices to collect information from users, leads to virtually unavoidable surveillance, and supports intensive profiling analytics, further eroding individual rights to privacy and threatening the fourteenth amendment rights of equal treatment under the law.

The micro-social concern is how individual choices and decision-making around digital and face-to-face behavior have implications that individual social actors may not fully recognize. At a more macro-level, the question becomes: How do the realities of digital-space environments reflect the interests, control strategies, and power relations of social institutions in view of countervailing attempts to protect or elevate individual social agency? This is also addressed in consideration of major social institutions making choices and decisions that affect individuals in ways that individuals may not fully recognize. Together, these chapters demonstrate the importance of continuing to examine the broader implications of how we interface ourselves as physically present and digitally present social actors, what it is to be *mindfully* co-present, and how we locate ourselves in the struggle between social control and social agency.

Each chapter in this book has carved out one focus among many in order to address one or more aspects of the micro-social and macro-social concerns integral to digital life. This has not been undertaken as a criticism of the co-present individual. It has not been undertaken to bemoan the loss of physical presence centrality. Rather, authors have attempted to provide a relatively balanced approach at the micro-social level by illustrating an understanding of the draw of digital life, the importance of co-presence, and the positive role of digital engagement for individuals, relationships, and social integration. Similarly, the macro-social foci in this book are not rooted in a wholesale condemnation of the Internet as an instrument of social control and power abuse. Chapters also show a balanced approach at the macro-level. For example, the historical use and continued potential for successful digital activism is recounted; and the power of individuals to collectively redefine "piracy" and ownership is demonstrated. Balance in the face of reasoned and informed decision-making and practice is, in fact, what authors see as the hallmark of this collection of works.

Did the Authors Succeed?

Our greatest success would be to help inspire two outcomes. First, that we motivate the intensively co-present readers—particularly the Millennials and post-Millennials who have grown up as digital natives, without a physical-presence-only comparison—to take a step back and view their digital engagement with greater complexity and detachment. Ideally, this would trigger introspection over their co-presence needs and choices. Second, that for those who are more physically present, we evoke a more layered understanding of what motivates and escalates intensity levels in the multi-platform connectivity of others. In this regard, our interest is to acknowledge, but move beyond, generation-wide attributions of narcissism and addiction. We feel it is important to identify prosocial factors such as social-psychological well-being, digital community, agency in the competition of values, and feelings of social integration. To do so, we join those who sidestep the either/or approach often implicit in work critical of extensive digital life.

What Other Types of Awareness Came with Digital Abstinence?

Millennials who forwent digital presence and lived for twelve hours as only physically present among the co-present population provided valued insights that underpin this reader, as well as specifically informing six of the chapters. Their journals illustrate the essential and internalized nature of millennial digital presence in their co-present lives—insights that may be more broadly applied as similar levels of digital intensity are developing among Baby-Boomers and post-Millennials. For example, authors demonstrate how difficult it is for digital natives to use the *sociological imagination* adequately when it comes to living outside familiar routines of co-presence. In order to obtain sufficient detachment to seriously consider the implications of digital life choices, abstaining either literally, or vicariously, is a critical piece. Hopefully, readers can identify with digital abstinence participants in order to share a raised awareness about the effect of digital life choices.

Despite the number who reported intense cravings (65%) and feelings of withdrawal or addiction (24%), as well as those who failed to remain abstinent and awake for all twelve hours (31%), not all reactions were purely negative. Even these participants recognized a variety of positive aspects to resisting overly dominant co-presence. The goal and overall narrative of this book are supported by the abstinence participants' confessions.

Recognized Need to Rebalance Co-Presence When in Shared Physical Presence

> It makes me value more the time we should spend with people that we have face-to-face interaction with. In my situation, the time I spend with my kids is more valuable than any post from a "friend" on Facebook. The interaction I had with my mother is more valuable than seeing a picture of what a "friend" is eating. It is important to set boundaries to

> what extent technology will affect my life, and think about what is really important.
>
> Undoubtedly, the millennial era has allowed more accessibility to technology, but finding the balance whether accessibility of technology does not dictate my life is now a challenge for me. Because I am fully aware of this, I can take the appropriate steps to appreciate my loved ones more, slowly change my overall urges … .

These quotes represent the participants (39%) whose journals include references to recognizing the value of limiting co-presence when sharing physical presence with others.

Recognized Formerly Unnoted Qualities of Physical Presence

> Having the opportunity to look ***at*** the world instead of looking ***for*** the world gave me time to process the world around me.
>
> Not being connected to the world that day gave me a chance to just enjoy the simpler things in life. Going for the hike and really taking in nature, as one should, was a lot of fun.

Others described noticing the gardens and beauty in their own neighborhoods for the first time, or actually hearing what the ocean waves sounded like without being accompanied by ear-bud-fed music. Participants acknowledged the value of re-balancing their co-presence, being pleasantly surprised (12%) or happy (27%) about a physical presence experience, though neither of these moods remained consistent throughout the half-day.

Time Management and Focus

> During the next hour I got a lot of reading done and felt content about myself because I read a whole chapter within an hour which would have taken me longer if I had the phone right next to me. I knew at that moment music was not part of my distraction—-it was the constant messages sent to me by my friends.
>
> One good thing that I did notice is that I get things done much faster without any of these distractions. I did my usual morning routine in 45 minutes instead of 1 hour and a half. I also felt like time was going so much slower than usual, and I began to think that what I had planned for the day was not going to be enough.
>
> Usually while I study, I listen to music or have the TV on, but I didn't so it was unusually quiet in my room for a change. It did help with minimizing distractions such as

> texting or going on the Internet, so I was able to focus more on what I needed to.

These passages represent the two out of every eleven participants who became aware of the threats to their study focus that are posed by their usual co-presence, and the 29 percent who recognize feeling that they waste too much time in digital activities that hold little actual meaning for them.

Freedom from Response Pressure

As mentioned earlier in this book, studies have found that here is a tendency for those who are engaged in frequent texting to continue to escalate their expectations of number and immediacy of responses. For many, this builds to a point where the messaging—once experienced as bonding—shifts to feeling burdensome, and reduces satisfaction levels in the relationship. This dynamic may explain why near one-quarter of participants reported feeling relaxed (22%) at some point (though this, too, did not endure for the full twelve hours).

> I felt at peace. I didn't have to worry about explaining that I hadn't written back to a text because I was walking my dog. I really enjoyed the feeling of not being tied down to a device.
>
> Being disconnected from the technology world was a great. No worries and stress about emails.
>
> I realized something that I was not able to do was text any of my friends or family members. Without texting, I felt a relief and didn't feel worried or stressed out ... Not being able to use my phone made me feel relaxed.

Freedom from Constant Social Judgment

> I probably go on Facebook and Instagram at least 10 times a day. I am constantly viewing people's profiles and comparing them to my life. This constant access to people's life acts as a non-stop reinforcement of values, pressures, and mass media. Although it was stressful going a day without social media, it was a relief to just be myself for a day and not post what I am doing in order to receive praise through how many likes I can receive.

Tuning Back in and Venturing Away from the Digital Shield

About one in every five participants (19%) reported using their phones to shield themselves from potential social awkwardness. However, many of their journal entries revealed this to be a form of self-overprotection that deterred them from developing confidence in their social skills.

> I know I am a shy person and though I recognized that I turned to my cell phone to deal awkward situations, I find that I warm up quicker to people I don't know without my cell phone.
>
> In order for me to avoid that feeling of being uncomfortable, I would turn to my phone ... to distract me from the situation at hand. However, not being able to turn to my phone and face that inevitable meeting and uncomfortable feeling head on, I built confidence in myself. I know I can face those situations and enjoy what I am doing with whoever I am with instead of trying to avoid it. All in all, this assignment had me feeling more confident in myself and feeling more involved with the people around me.

Others realized they avoided time alone with their thoughts, becoming aware that introspection is a necessary part of psychological health.

> The inability to zone out but instead being distracted by everything that was going on around me was a bit difficult at first ... I decided to focus on myself and get lost in my own thoughts. By the end of the hour work out I really appreciated this time and was able to think about life, problems, issues, and myself. It was in a sense therapeutic.

Revisiting Limited Periods of Physical-Only Presence

Although a significant number of participants (29%) report having a greater respect for digital life, fully one-third (34%) stated the importance of everyone being more aware of their co-presence balance and making more mindful digital life choices. Further, a small number (8%) actually reported their intentions to establish some temporary periods of physical-only presence in the future.

> My overall experience during the twelve hours was something **I would definitely do again** without it having to be an assignment. I liked the idea of not having to worry about who is trying to contact you or who is following my new follower or who posted a new video on YouTube.
>
> The significance of the assignment went beyond having to live one day without any chip based technology; it opened my mind to consider in what way I was living my life. It made me realize that in order to not completely devoid myself of being a human it is

> okay **to sometimes turn off the phone for one hour** and enjoy simply being alive.
>
> I realized that most of the times when I would go into my Facebook I would scroll down to look for pictures. Most of the pictures that were posted were already on other social media, like Instagram and I had already viewed them. I realized that I would waste more time on all of them than using my time to do things around the house and spend more time with my daughter. **I have actually decided that Sundays will be the day where I will limit my technology use** and will make it a day spent with my daughter, family and friends.
>
> Although I see them often I rarely spend time with them and if I do I am usually on my phone while they watch TV. This time the difference in our interaction was huge. I completely turned my phone off and just played with them. I taught them the folk games we were taught growing up and the look in their face was priceless. I had not taken the time to really get to know them and be more present when I am around. I was able to see that **I need to put my phone down more often and live in the moment** and enjoy life as it is happening in front of me.

It is important to realize that all of these participants are describing the need to consciously detach, even briefly, in order to maintain the insights they gained. Their responses underscore authors' understanding that self-reflection about internalized and routine behaviors is extremely difficult without a mechanism for detaching oneself from those behaviors. For these participants, building a detachment routine into their actively engaged digital lives was a means of maintaining power over digital connectivity and its place in their co-present lives.

Foci and theoretical lenses vary from chapter to chapter; however, chapters share a sense of primacy in promoting informed decision-making and self-awareness wherein individuals actively consider the balance and interpenetrating aspects of their physical presence and digital presence, as well as better understanding the broader implications of living in a digital era. Ultimately, the goal across chapters is a thought-provoking gestalt that underscores a need for greater self-awareness—one in which individuals bring more conscious reflection to their digitally-mediated choices—one in which they better understand the visibility of their online presence and its impact. This book presents a multi-faceted, empirically supported and theoretically-anchored collection of original works demonstrating the importance of continuing to examine the broader implications of how we interface ourselves as physically and digitally co-present social actors, and how we address our social agency.

15

ANNOTATED METHODOLOGY

Investigative Process: Research Reflections

Cristina Bodinger-deUriarte

Overview

All authors included in this book provided original, empirically-based work exploring digital life. Each chapter makes knowledge claims about the nature of digital life and human experience. Whether these knowledge claims hold sway depends, in part, on the reasoned argument and evidence provided. However, the way such evidence was obtained is also significant to the credibility of the knowledge claims. Ultimately, "understanding and reporting how or why people behave as they do involves analyzing and presenting reality. In practice, this means sharing with an audience a convincing account of what was observed and its meaning" (Marvasti 2011:4). What follows is a description and commentary on how these authors collected evidence that enables the readers to believe the authors have, indeed, provided "a convincing account of what was observed and its meaning." This chapter looks at commonalities underlying author investigations of the empirical realities of digital life. On one level this focuses on the overall outlook informing the research process. On another level this addresses different empirical approaches and methodological tools among the chapters.

Outlooks Informing the Research Process

Chapters contributing to this book most often take constructionist and phenomenological approaches, root empirical observations in social-historical context, and rely on grounded and emergent processes. Constructionism and phenomenology affect the framing of the research questions, which in turn, affects the way in which evidence is collected.

Constructionist Outlook

There are three distinctive research elements within the constructionist framework:

1. Subjective ideas from research subjects are part of the overall information considered when developing knowledge claims about human experience.
2. The focus is not on developing universal and enduring explanations and predictions, but on how the meaning, practices, and outcomes are developed in particular contexts.
3. Constructionist researchers are aware of the relationship between research findings and public understandings that can act together to support a particular politicized or social change stance.

Constructionist research acknowledges that how subjects interpret their actions has an impact on the choices they make about such actions. "Social constructionism and qualitative research is a natural marriage, wedded by a mutual respect for the complexities of the human experience" (Roller 2015). To truly understand the patterns of behavior, at the micro-social level, requires one to understand patterns of shared motivation or shared meaning attributed to the behavior, and whether or not alternate behaviors are seen as reasonable and fitting. This does not imply, however, that choices are made without limiting factors. The contexts in which behavior takes place are, in part, defined by the limits that constitute what people see as normative. In other words, socialized concepts about what is *appropriate and expected* help to limit the set of behaviors individuals see as reasonable alternatives. It is the push and pull of unfolding and changing social expectations, and consequent possible actions, that affect shifts in practices. Finally, part of what comes to be understood as possible, reasonable, and normative, ties back to developing knowledge about the consequences of common practices, and the choices made by various social groups or cohorts. Ultimately, this outlook is predicated on the idea that groups of individuals can, through social agency, change practices through informed choice, even in the face of limiting contextual traditions and larger social forces.

Phenomenological Outlook

Constructionism and phenomenology align, to some degree, across the three elements enumerated above. Briefly, phenomenological work is also sensitive to: (1) the role of subjective human experience in contributing to the research, (2) the need to attend to the shifting or variable contexts within which social phenomena develop, and (3) the potential policy implications of their findings. Phenomenology, by drawing on the specific views of those engaged in the phenomenon—in terms of life as it is lived—provides a natural comparison to meanings developed by those outside the phenomenon. For example, digital natives may have significantly different motives for their digital intensity than the explanations about them provided by older cohorts. In this way, phenomenology is specifically concerned with how experience compares to taken-for-granted claims. As Lester explains:

> Phenomenological methods are particularly effective at bringing to the fore the experiences and perceptions of individuals from their own perspectives, and therefore at challenging structural or normative assumptions … enabling it to be used … to inform, support or challenge policy and action (1999:1).

Although constructionists and phenomenologists share this as a hallmark, not all phenomenological work embraces the constructionist view typical of this book—some is much more deterministic in its orientation. As Introna noted:

> There is not a unified phenomenological tradition or approach … The phenomenological tradition consists of many different approaches that share certain characteristics … phenomenology suggests that there is a co-constitutive relationship between us and the phenomena we encounter in our engagement with the world. In this sense phenomenologists would suggest that to understand the technology/society relationship we need to reveal how they co-constitute each other—i.e. draw on each other for their ongoing meaning and sense (2017).

In other words, the phenomenological approach that underpins behavioral and social science research methods in this book considers how changing practices of individuals and cohorts shape technology, and thereby the realities of digital life (constructionist), versus how broader social forces and developing technology shape the realities of digital life for individuals and cohorts (determinist).

Because digital life is the broad focus of the chapters in this book, the methodology is greatly affected by the rapid changes and dynamism of digital experience. This explains, in part, why most of the works include some treatment of the social historical roots of the phenomenon of interest. Addressing socio-historical circumstances fits well with the phenomenological approach that looks at knowledge claims as contextualized and situational, rather than enduring and universal.

The combination of a social historical context, the phenomenological perspective, and an inductive reasoning process that may generate grounded theory, is typical of the work in this book that "begins with the empirical world and builds an inductive understanding of it as events unfold and knowledge accrues" (Charmaz 2008:155). The key methodological element is the grounding in life as it is lived, however.

The Grounded Theory Approach

Finally, it is important to consider how theoretical frameworks or concepts are incorporated into the investigation of a phenomenon. In several chapters, theoretical notions underpin the focus without being formally articulated. In others, they are made an explicit part of the explanatory narrative. The role of theory in research varies. Most often it provides a springboard, particularly in quantitative work, where theoretical precepts are tested through specific hypothesis in a deductive process. Less typical are research projects that use a more inductive reasoning process.

- Deductive Process: theories → hypotheses → data collection / empirical investigation → confirmation / alternative theoretical proposition.
- Inductive Reasoning: data collection / empirical investigation → pattern identification → analytic supposition → explanatory narrative / grounded theory.

The chapters in this book primarily present examples of inductive reasoning processes that started from general observation, rather than a specified research question. The outcome of this process can be grounded theory.

> Grounded theory fosters openness to what is happening in the empirical world. That means studying data and developing an analysis from conceptualizing these data rather than imposing a theoretical framework on them (Charmaz 2008:163).

Whether the research is exploratory or follows-up on well-established work, the aim of both deductive and inductive approaches is to make defensible knowledge claims that include ties to why the findings matter—to provide a "so what" factor. Frequently this is the point at which findings are tied to either known theoretical concepts or newly-generated explanations. In other cases, the ties are elaborated more as verified patterns of behavior and observable social consequences, as is often the case with grounded theory using emergent research methods.

The Methodologies

The works collected are primarily qualitative in nature, which is more typical of constructionist and phenomenological approaches. Boundaries can be blurred, however, due to some overlap when highly-qualitative material, like the journal entries, are systematically coded and analyzed for correlations. Content analysis (CA) also bridges the qualitative/quantitative divide, somewhat. CA distributes qualitative material into pre-determined analytic categories and then, using descriptive statistics, generates knowledge claims based on relative frequencies and comparisons among the categories. So rather than think of qualitative and quantitative approaches as a simple binary, an "either or," it is more fruitful to think of it as a continuum. At one end of the methodology continuum in this book is the phenomenological and qualitative grounded theory approach. At the other end is the more quantitative survey method using standardized closed-ended and forced-choice items based on an established research question.

The empirical basis for data collection and phenomenological investigation are detailed and discussed for each chapter. This is organized by the primary method involved, rather than by chapter. The goal is not to critique the specifics of research undertaken in individual articles, but to consider the value of the varied methodological approaches using authors' research as concrete examples.

Social History

Social historical context is provided for aspects of the subject addressed in most chapters, regardless of the primary method used. This is to be expected because social history is part of constructionist and phenomenological process, as well as an acknowledgment of the rapid changes in the digital world. In effect, this is an emergent research approach. Emergent research acknowledges that "the present arises from the past but has new properties. These novel elements of emergence distinguish the present from the past and make it distinctive" (Charmaz 2008:155). Consequently, to appreciate realities of digital life in the present, requires some reaching back to track critical digital developments. How far one "reaches back," to what degree, and in how much detail one "tracks critical developments" is part of what distinguishes phenomenological *context* from papers in which the social history is the *primary* means of explaining the current dynamics of the phenomenon and its emerging impact on life as it is lived.

Phenomenological Social History

Chapter 12, "Power and Money, Explaining the Rise of Digital Media through Surveillance Capitalism," provides an excellent example of using phenomenological social history as a defining methodological element. Choi addresses macro-social consequences of digital life by taking a broader social historical means of illuminating the complex unfolding phenomenon. Choi tracks critical incidents contributing to technological change, and to the eventual development of the Internet in its current form. His methodology may best be described as emergent grounded theory and phenomenological empiricism through social history.

The information gathering process involves scholarly materials, evidentiary reports, official documents, the purposive tracking of the rise of influential organizations, the impact of shifts in policy, the ramification of decision-making among powerful figures, and the societal context and vested interests surrounding the development of Internet technology. Choi's materials include directly relevant, declassified government documents and Senate subcommittee reports. This results in a deeply informed social history rooted in macro-level social forces and their implications for members of society. This phenomenological social history generates a complex picture of the realities of surveillance capitalism that is embedded in, and shapes priorities of, digital environments. The resulting, grounded theory provides an explanatory frame drawing from power elite and conflict theory.

The challenge in executing the social-historical, inductive, phenomenological research of this type is the degree to which it relies on the diligence and informed background of the researcher, as well as access to pertinent documents. Knowing which paths to follow in locating reports, and tracking relevant changes in law and institutional policy, depends on deeply cultivated background knowledge. The limitations of this approach, therefore, reflect the limitations of the researcher's preparedness.

Phenomenology and Digital Auto-Ethnography

Chapter 13, "Knowing You Better than You Know Yourself: Manufacturing Perceptions," also takes a phenomenological approach, but uses auto-ethnography and digital participant observation fieldwork as primary methods. A post-modern, emergent method, auto-ethnography has been described as "an intriguing and promising qualitative method that offers a way of giving voice to personal experience for the purpose of extending sociological understanding" (Wall 2008:38). As indicated by the term, three elements combine in this method: the self (auto), the socio-cultural connectivity (ethno) and the research effort. Different studies emphasize different elements. Nitzani emphasizes the socio-cultural connectivity, then balances the self and the research effort (existing scholarship, and pseudo-experimental intervention) to "extend sociological understanding." His use of field notes and consideration of social artifacts (such as commercial messaging) mirrors traditional ethnographic process. Nitzani's focus is the dialectic relationship between individual digital space activity, and macro-social commercial interests.

Nitzani treats the various digital platforms he frequents as the "field" in which he is a participant observer. Initial digital field experiences led to an overall research interest in data-driven financial processes that influence pop-up digital messages, the import of this commercial messaging, and why it might have an impact on users who did not seek out such messages. Further experience-based observations were used to inform a thematic framework to organize subsequent observations, and to indicate topics requiring scholarly investigation through existing literature.

- Deceptive communication practices
- Propagandistic and manipulative imagery
- Overt capitalistic interests
- Algorithm-based constructions of reality.

Digital field experience, contextualized through scholarly materials, yielded a grounded theory of the commodification of digital life, and the role of big data in shaping the perceptions and, consequently, the decision-making of users. Nitzani's own raised awareness of his digital environment, led him to employ pseudo-experimental emergent methods. Nitzani purposively altered his own profile information and search topics to test whether algorithmic responsiveness would lead to changes in commercial messages and apparent changes in filter-bubble perimeters. His results led to empirically-based suppositions on potential ways users might unbalance the algorithmic construction of a commodified reality. This contributes to the discussion concerning the limitations of individual agency within the big-data influenced context of digital life. As such it emphasizes the constructionist and phenomenological interest in supporting social action or policy change.

All participant observers and ethnographers situated in familiar settings know the value of insider knowledge paired with scholarly distance gained through methodological practice. However they share the same pitfalls, as well. One of the primary criticisms is the degree of subjectivity and the absence of quantifiable material. It should be remembered, however, that when the intention of the research is more exploratory, descriptive, and evocative, such critique might be countered by positioning observations within established literature that supports different aspects of one's knowledge claims. Ultimately, an auto-ethnographer's self-reflection should ask whether I "offered something new that is intended to help others understand the world of which I speak, made claims that I can legitimately make, presented a complex and nuanced story, and promoted dialogue" (Wall 2008:48).

Time-Interval Diary Study

The following describes the methodology used by the five colleague research team (Apardian, Bodinger-deUriarte, Hryniszak, Okamura, and Valgeirsson) who authored six of the chapters in this book:

- Chapter 1: Offline as Misaligned: Millennials Coping with the Loss of Digital Presence
- Chapter 3: Powering Down: Theoretical Lenses to Examine the Agency of Our Smartphones
- Chapter 5: Digital Dependency Interrupted: Profiles of Withdrawal for Self-Described Internet Addicts
- Chapter 6: How Do I Love Thee? Let Me Text the Ways: Interfacing Intimacy
- Chapter 7: La Familia in Digital Space and Face-to-Face: Millennial Latinx Navigating, and Reconfiguring Conceptions of Familismo
- Chapter 8: Is Unmediated More? When Physical Presence Does Not Equate to Digital Presence

Each of the above chapters draws from diary studies methodology. The variety of subjects supported by the journal data testify to the richness of diary-based material. It is also an indication of how much can be gained by accessing and repurposing secondary data sets, even if data were gathered for a different reason.

In diary studies, data is provided by the frequent entries participants write about what they experience in their daily lives. These self-reports catch descriptions and reactions to events in a way that is not possible using other traditional methods. The immediacy prevents memory erosion, and the frequency works against over summarizing the event as a whole. Further, changes in responses to an ongoing event are more easily tracked when self-reported affect is periodic, instead of a one-time reflection. As explained by Bolger and colleagues, "diaries, self-report instruments used repeatedly to examine ongoing experiences, offer the opportunity to investigate social, psychological, and physiological processes, within everyday situations (Bolger, Davis, and Rafaeli 2003:581). Another benefit is that without fieldworkers or interviewers, the context of the phenomenon remains unaffected. In other words, "A fundamental benefit of diary methods is that they permit the examination of reported events and experiences in their natural, spontaneous context" (Bolger, Davis, and Rafaeli 2003:581).

Diary analysis can be conducted on archived material, or on self-reports specifically designed to address a particular research question. Material for diary analysis can be drawn from individual personal diaries; a set of one-time reflections; or monthly, weekly, daily or even (as in this case) hourly journal or log entries. Any set of narratives that communicate personal experiences, affect, and reflection, in a time series of some kind, are considered *time interval diaries*. This narrative material can yield rich data approachable from multiple angles. Social and behavioral science or communications researchers "analyze narrative data in order to develop an understanding of the meanings our participants give to their surroundings, to their lives, and to their lived experience" (Kim 2016:189).

For the six above-mentioned chapters, the narratives comprising the raw data resulted from an assignment in four different sociology courses; the topics were: socialization and adult life; qualitative methods; popular culture; and introduction to sociology. The digital abstinence journaling was part of a student-centered, active-learning project about how cohort-based experiences can differ in significant ways. It was intended to promote deeper perspective-taking when discussing older cohorts. Almost 800 such journals had been archived by professors. The assignments were not initially meant to provide research data. The research team recognized these journal assignments as narratives ripe for diary analysis, however.

The team chose only those journals written by millennial students (aged 18–34, but primarily under 27) in the more recent years. Aside from some simple demographic information, the research team received no identifying information on the individuals. This yielded 420 digital abstinence journals. Digital abstinence journals were comprised of hourly entries for twelve consecutive waking hours. Individual participants chose their start time and date (within a multi-week period). Participants gave up all use of smartphones, computers, tablets, and other digital media access devices, music streaming and MP3 players, etc. At the end of each hour of the twelve hours, participants described:

- What they would normally have been doing, but gave up to remain offline.
- What they did instead.
- How they felt during that hour.
- Whether their offline status had an impact on anyone else during that hour.

Most participants also wrote a substantial final reflection after the twelve hours ended. There were no other journal requirements, including length of entries, so participants were free to express whatever they wished.

The research team recognized the value of these assignments as raw data. The team was interested in what participant reaction to digital abstinence could reveal about the norms and expectations around digital presence, and how that would manifest in terms of moods and behavior. The fact that students had been directed to write in one-hour intervals was quite fortunate for the later research effort. This enabled researchers to ascertain mood changes throughout the experience, a type of data that time-interval diaries are well suited to provide. In fact, according to Bolger and colleagues, "Most diary studies, particularly those focused on within-person processes, involve time-based designs" and "Hourly variations in specific processes or experiences…[reveal] within-day fluctuations of moods" (Bolger, Davis, and Rafaeli 2003:589).

For many of the participants, it was quite literally the longest they had been without their smartphones since they acquired them as youngsters, and the first time they had been required to navigate a substantial part of a day without any active digital presence. This constitutes a rare and significant twelve-hour event for participants. The journal entries, therefore, provide appropriate data because "an event-based design is usually most appropriate for diary studies of specific classes of phenomena or processes, especially those that are isolated and/or rare" (Bolger, Davis, and Rafaeli 2003:591).

Diary analysis follows processes akin to those used to code open-ended interview responses, field notes, and content analysis materials. The research team worked together to develop a coding scheme after reading through dozens of journals. Coding was an iterative activity in which codes were added to the content analysis instruments throughout the process, as needed and agreed upon by the authors. Already-coded material was then re-coded to address any new codes added. Ultimately, each journal was coded with the standardized instrument comprised of (among other things), twenty-three emotional response codes, twenty-three codes concerning self-reflection, eight codes about impact on others, and nine additional participant outcomes codes. The data were entered into SPSS enabling frequency and cross-tabular analyses. Participant narratives are also quoted by way of illustration and to provide deeper understanding of participant experience, and the meanings they attributed to their subsequent reactions.

Potential Limitations

As mentioned above, participants chose any date that suited them within a multi-week time frame; they also chose any twelve consecutive hours of that date. This was an attempt to avoid days or times in which particular work or course assignments, or non-typical events, might skew the ability of participants to comply. Many chose to do a couple hours of emailing and social media use immediately prior to their twelve hours of abstinence, while others chose to wake up without their phone alarms in order to start from the very beginning of the day. Again, this was to enable participants to give themselves the time period least likely to work against compliance.

One limitation in self-report data is the possibility of inaccuracy due to the desire to present oneself in a better light (*social desirability* bias). There are a variety of ways this can be addressed—most often through anonymity—but the concern here was that these were assignments turned in to a professor for grading. If compliance affected the grade, one would expect participants to be tempted to fake compliance, even when they had not fully abstained. However, students were assured that their grades did not depend on their abstinence success, but on the communication of their experiences. They were told that if they absolutely could not mange to abstain for the full twelve hours, to be honest about the interruption, and then return to the experience and continue to do their best to complete the rest of the twelve hours. The goal of a self-informed learning experience was emphasized. This increased the research team's confidence in the accuracy of the self-report data because the pressure to fake-it for a grade had been removed. Further, confidence in the data rests on the fact that almost one-third of participants *reported* being unable to comply, yet *resumed* the effort after briefly violating their abstinence.

Another limitation (as with any narrative analysis) is the possibility of misinterpreting the meaning of written passages. Methodological checks need to be built in to avoid this as much as possible. Codebooks were detailed and included multiple illustrative quotes. Training was extensive. Several journals were coded in common during training, and at two different interim points to test for inter-coder reliability. Inter-coder reliability is defined as the degree to which those coding the same journal recorded the same codes for the same passages in the entries as one another. Weekly meetings in which all coding questions were flagged for group decisions aided consistency. Further, the iterative coding and re-coding process made it easier to catch any overlooked codes from the first pass through. The size and multicultural membership of the research team that, nonetheless, achieved high inner-coder reliability adds to the confidence that interpretive bias is minimal.

Content Analysis

Formal content analysis (CA) is a highly-adaptive method that involves developing a standardized set of categories. Material is coded so that it can be distributed among these categories in a consistent and systematic way. Numerous types of materials are suitable for CA. As explained by Chandler (nd):

> Content analysis means analyzing documents or processes for information that may be used to find out more about a situation, person, or organization. These documents are … grouped into five types: written; oral; iconic; audio-visual; and hypertexts.

Four of these five types of CA are represented in this book. Advice column letters are written documents analyzed to discover more about conflicts triggered by digital space activity (Chapter 9). Iconic CA involves pictorial, photographic, and other static images such as the photos of tattooed individuals in print and uploaded to websites (Chapter 2 and Chapter 4). Audio-visual CA may involve videos, television, films, etc., such as the uploaded tattoo videos also treated in Chapter 2, and blog-situated material in Chapter 4. Hypertext-based CA refers to any material that may be found on the Internet. In this case, examples are provided by Apardian's typology of activist websites (Chapter 11) and Sloan's analysis of the communicative intent of blogs, gifs and memes within digital LGBTQ+ communities (Chapter 4). Consider that both Burns and Sloan also used interactive material in their CA which adds a layer of complexity:

> In order to content analyze interactive media content, it's important to understand that interactive media users are more than just receivers or consumers, as they were with earlier media. They have an active role in adapting, altering, and even producing content (Skalski, Neuendorf, and Cajigas 2017:202).

Content analysis has proven to be an exceptionally-versatile method, particularly in turning qualitative material into quantifiable categories that can be compared by relative frequencies and correlations. This translation of qualitative material into well-defined categories that can then be counted, enables comparisons across time, settings, and genres. This flexibility is applied by Burns to track changes in the tattoo culture's treatment of tattooed women's bodies from tattoo print magazines, to tattoo-culture reality television, to tattoo-dedicated websites (Chapter 2).

Although the process is similar to diary studies, CA material may have an even greater need of operational definitions due to interpretive variance across image and performance-based material. There is also a need to identify exactly which elements are included and which are omitted. As the meanings of single words, phrases, or images may be shifted in view of the context, a process that retains important knowledge of contexts can be critical. As in diary studies, detailed codebooks, clear and distinct categories, substantial training, and the assurance of inter-coder reliability are essential to confidence in the findings. For a more detailed step-by-step coding discussion MacNamara's 2005 article is recommended as a concise and practical guide. Meanwhile briefly *summarized aspects* of author process and rationales are provided below to illustrate various aspects of CA, its strengths, concerns, and limitations.

Chapter 2: Digital Ink: Social Media and Tattoo Culture in Consideration of Gender

Burns conducted a content analysis of photos and videos uploaded to the *Inked* magazine Facebook (FB) pages, and its culture tattoo-dedicated websites. The analysis covered the *Inked* magazine pages site from 2011 to 2017. The only included material consisted of photos and videos posted by *Inked* magazine and photos and videos posted by their followers or audience on the magazine FB and linked sites. All other posting on the FB pages site (such as advertisements and click-bait) were omitted as not directly pertinent to the research question, which included *self*-objectification.

A content analysis tally sheet was developed to note the relative frequency of presentational elements by gender and ethnicity, among other things. Because the tally sheet is a measurement tool that standardizes the definition and categorization of images, it diminishes the bias of non-standardized analysis. CA categories included: art focus, titillation focus, modest clothing, revealing clothing, lingerie and underwear, nude neutral, nude sexualized, mainstream beauty, and alternative beauty. Burns included a count of fairly straightforward elements such as age and body-type, as well as more interpretive elements such as artistic focus and titillation focus. Operationalizing terms is critical when there is a wide interpretive margin. The following is an example of operationalization from Burns' work:

> The category of "nude neutral" refers to an individual who poses nude in these tattoo magazines, but in a non-sexual way. In this case, sexualized locations are covered up and individuals are posed or modeling in a neutral stance (standing up with their hands to their sides or holding a cloth or towel that covers sexualized locations) which allows the focus to be *only* on their extensive tattoo work.

Finally, in order to avoid researcher bias, Burns used a research assistant, detailed training sessions, and debriefing meetings as part of the process. The inter-coder reliability rate was very high (94%), increasing confidence in the findings.

Chapter 9: Interfacing Conflict: Advice Columns and Digital Life

This is a relatively simple and straightforward example of content analysis. In this case, Valgeirsson and Bodinger-deUriarte read through numerous advice columns and discussed the increasing appearance of conflicts triggered by digital-space statements, posts, or behaviors. This led Bodinger-deUriarte to develop a framework for conducting CA, focused on which digital issues were being addressed in advice columns and to what degree. This was an iterative process with a trial coding period and then revision of the CA instrument. The final codes categorized digital conflict letters by these primary issues: (1) boundary violations using social media; (2) cyberbullying; (3) friending/unfriending; (4) netiquette; (5) online dating; (6) outcome anxiety; (7) relationship conflicts involving digital behavior that is essentially the same as non-digital behavior; and (8) miscellaneous other digitally-based or triggered conflict.

Santhoff developed the social-historical to present-day overview. This put advice columns into a broad context that explained the social significance of advice letters and underlay which columns were selected for CA. Santhoff chose: Tell Me About It/Carolyn Hax (established 1997),

Dear Abby (re-established in 2002 when taken over by Jeanne Phillips), and Ask Amy/Amy Dickinson (established in 2003).

Content analysis is often performed on materials *sampled* from extremely large potential data pools. Selection criteria need to be carefully established because characteristics that differ between the included material, versus the excluded material, potentially have great impact on findings. Sampling should directly connect to the research purpose. Bodinger-deUriarte, Valgeirsson, and Santhoff sought to establish the types of digital conflict concerning individuals, and if these conflicts reflect rising social concerns about digital life in general.

In addition to being well-established (verifying their legitimacy as sought-after advisors), several criteria led to the column selection. First, they needed to be general, all-purpose advisors. Columns with a narrow or specialty focus, like business or sexuality, would provide limited representation of the variety of digital conflict issues. Second, columns needed to pre-exist online competitors. Columns that were first established online might have developed more Internet-active audiences, or might attract disproportionately higher levels of digital conflict questions simply *because* the columns are online.

The search for digital-conflict letters was initially problematic. Although accessibility was not a serious problem, reading through thousands of letters was not practical and was beyond our time frame. The initial plan was to conduct key term searches for the most recent few years of online archived letters for these three advice columns. Key terms included, among others: online, Internet, Facebook, social media, post(s), and tag(s). However, not all archives had sophisticated searches that responded to key terms. Also, the sheer number of terms to search by became unwieldy, so this process was abandoned. We tried searching by title, yet not all letters had the digital issue represented in the titles—also, titles were sometimes revised from one archive to another, meaning one had to remain alert to avoid duplicate letters in the pool. The key word search produced 109 letters. Title searches added 52 letters. After that, Bodinger-deUriarte scrolled through the following archives, scanning non-selected letters for any digital referents: *Uexpress*, ArcaMax Publishing, *Seattle PI, Denver Post*, and *Washington Post.* This process added another 94 for a total sample of 255 letters.

Although the categorization process was simple, obtaining the sample was extremely tedious. Ideally, research team members would each have divided responsibility by columnist, and read through each letter in order to be certain no relevant letter was overlooked.

Chapter 11: Community or Catharsis? Online Activism, Digital Community, and Social Agency

Apardian's work is largely phenomenological and is rooted in social historical trends in physical space activism. He also applies well-known theoretical precepts that he extends to digital life, such as social capital and the loss of localized community. Consequently, his CA is designed to be more deductive and less about grounded, discovery-oriented induction. Apardian uses content analysis differently than it is used in other chapters, because it is not the foundational material. Content analysis, in this case, is meant to provide a snapshot that illustrates the theoretical propositions primary to his argument. The descriptive intent is manifested in the outcome, which is a typology of eighty-five activism-dedicated websites. A typology consists of conceptual categories that are only intended to label and describe. Apardian generated six categories according to what the websites offer to activists and potential activists:

- Website is educational and awareness-oriented, in that it includes information on one cause or several different causes.
- Website offers a discussion board or other messaging for site members.
- Website encourages viewers to support causes through donations and provides a mechanism for processing donations.
- Website enables users to engage in electronic petitioning.
- Website facilitates volunteerism by enabling viewers to connect with volunteer opportunities.
- Website provides schedules and locations for events, gatherings, protests, and other physical space activist activity.

The above typology is roughly arranged from least to most actively involved in work to promote social change.

Adding this type of directional element to a descriptive typology, enables researchers to readily assess whether the relative frequencies among categories indicate an escalating or de-escalating phenomenon, or whether neither is significant. In this case, directionality enables the researcher to test predictions about activist websites manifesting greater or lessor support for committed activism. The set of categories that describe relative activism options are then tied back to the theoretical argument, as illustrations of potential activist participation. The resulting distribution is used to indicate how such sites foster or fail to foster actual political behaviors, and ultimately, whether online activism might dissipate, rather than facilitate, offline activism.

Chapter 4: From Back Stage to Digital Front Stage: Online Queer Community, Identity, and Emotion Management

Sloan's content analysis uses multiple types of material whose sampling is determined by digital platform, digital target group, and communicative purpose. Key words provide the raw data. Overarching thematic categories suggested by this data are used to organize data at this mid-point. Sloan further refines her analytic frame by developing conceptual sub categories in the final stage of analysis. This is significantly different than the other three

CA based studies: Burns's single level categorization of tattoo presentation across media; the digital conflict-area categorization across advice column writers (Bodinger-deUriarte, Valgeirsson and Santhoff); and the activism-level typology of targeted websites by Apardian. Each of these four CA designs is appropriate to the research focus and purpose—differences reflect research aims, not quality.

Sloan used Facebook, Twitter, Instagram, and Tumblr accounts for a four-month period to collect data from openly queer-identifying blogs, handles, and public pages. This generated 406 items, including text posts, photos, gifs (images in graphics interface format, or "moving images"), and memes, (discrete elements of knowledge, gossip, or jokes that are passed on culturally through social media). Content ranged from a few words in text, to short paragraphs, with most of the units containing images to emphasize the point being made.

Using an inductive approach, content was chosen based on key words or phrases including, but not limited to: gay, lesbian, gender, mental illness, symptoms, queer, femme, butch, masculine/feminine, misogyny, anxiety, depression, and dissociation. These items were then sorted based on latent content and four recurring themes emerged: identity formation, self-presentation, mental health, and community.

Multiple subcategories soon developed. Content relating to *identity formation* dealt with existentialism, nihilism, self-acceptance, and self-rejection. *Self-presentation* content contained themes of external validation, gender performance, and gender performance policing, and reclamation. The category of *mental health* broke down into issues of trauma, personal emotional growth, coping mechanisms, problematic behavior, anxiety and depression, self-diagnoses, and other mental illness specific concepts. Lastly, content relevant to *community* broke down into showing support for friends, partners, acquaintances, and strangers alike; general "shoutouts;" problems within the community; and one's place within the community.

One of the things that qualitative researchers in social and behavioral sciences wrestle with is controlling the degree of subjectivity in their work. Various aspects of this have already been addressed; however, there are other potentially-biasing sources. Sloan acknowledges the warning of methodologists that "Researchers who use an emergent method may simultaneously find that they are negotiating both an 'insider' and 'outsider' researcher identity" (Hesse-Biber and Leavy 2008:4). This is especially tricky when emergent and inductive methods are used by researchers who are studying *and* participating in hypertext environments. As Sloan says:

> One of the largest obstacles in my position as a researcher was in respect to the community in question. On one hand, I found myself wondering just how I attempt to learn about a group with which I personally identify? Does my identity as a queer, gender-

> fluid individual make me enough of an insider? … [Also] I had fully delved into Tumblr's microblogging community, acquiring nearly 10,000 followers … My outsider status also comes from my personal relationship with academia. I had to maintain a sense of personal detachment when analyzing content, as to not insert any personal bias into the content, while attempting to decode the manifest content based on what I know about the queer community.

It seems wise to conclude this section, with an observation Sloan makes. It pertains to all CA work, not just her own:

> Deciphering the ethics that accompany a research question can be just as difficult, if not more, than the data collection itself. Questioning my position as a researcher and just how the community in question should be represented, starts with the development of the research question and continues through to the end.

Standardized Self-Administered Survey

Only one of the chapters in this reader takes a quantitative approach, "Islands in the Stream: How Digital Music Piracy Became a Normal Activity" (Chapter 10). Data was collected using an anonymous, online survey questionnaire. The first few questions of the survey ask about age, gender, and ethnicity, to determine if there are any demographic effects when dealing with piracy. The next several ask about actual behaviors, to determine how people obtain music, what they do with it once they have it, and how they share it with friends and family. These are followed by a series of questions regarding subjects' attitudes toward the behaviors to assess whether or not they label these behaviors as deviant. Finally, subjects are asked about their perceptions of legality. Whether the behaviors in question are legal is not important; again, the key is to assess whether subjects agree with deviant labels or not. The research intent was to illustrate which gaps, if any, exist between attitude and behavior, and whether those attitudes and behaviors do or do not support the rationales behind the law.

All self-administered surveys risk using questions that do not match answers respondents want to provide, and losing data through skipped questions or, worse, corrupting data through inaccurate answer choices. Pilot testing surveys, and including a feedback mechanism, is an important part of the process. Okamura's pilot survey informed the revision of the survey used to collect the data. Two examples follow.

First, a few of the pilot's respondents provided feedback that the answer categories for one of the questions did not adequately represent their preferred answer. Okamura re-worded questions to be more inclusive, and provided options to select "other" with a blank to input text where provided. This allowed for the collection of rich data and to account for options missed when designing the survey.

Second, surveys may contain questions that respondents do not want to answer accurately because the question asks them about socially "undesirable" behavior. The pilot version of the survey was found to be heavily vulnerable to this type of social desirability bias. For example, the pilot survey asked respondents to indicate how much they agreed or disagreed with a statement like "It is OK to download music files from strangers without payment to the copyright holder." Such questions proved too direct and several respondents simply quit the survey during these sections. Therefore, the survey's focus was changed to emphasize behavior, asking participants about their devices, their activities, duration and frequency of listening, and how they share and receive music with their peer groups. Although the current version approached the issues differently—emphasizing the sociability of the behavior—the underlying theme of illegal activity makes the same bias possible. The decision to have respondents anonymously access the survey online was a strategy to help counterbalance this, one without a downside since the subject deals with online behavior.

Response rates can also be an issue for self-administered surveys. For Okamura this challenge was increased by the fact that the survey addressed illegal behavior. He met this challenge well by using multiple online recruitment strategies. Respondents were recruited via e-mail, and on Facebook and Twitter, inviting them to participate in an online survey using a supplied URL that led to a statement explaining the purpose of the research. Participants were asked to help create a snowball sample by forwarding the recruitment message to members of their e-mail contact lists and social networks. The e-mail recruitment consisted of eighty-three messages sent simultaneously; social network posts were automatically posted three times per day for two weeks by an application called Hootsuite. After one-week, a reminder was sent. Two further days of recruitment took place through the Web forum, Reddit, where a recruitment message was posted under the subReddit "r/SampleSize," a forum for posting and taking surveys. The survey was closed after 545 respondents participated.

References

Bolger, Niall, Angelina Davis, and Eshkol Rafaeli. 2003. "Diary Methods: Capturing Life as it is Lived." *Annual Review of Psychology* 54:579–616.

Chandler, Nick. nd. "Content Analysis to Determine Company Culture." Study.com. Retrieved July 25, 2018 (https://study.com/academy/lesson/using-content-analysis-to-determine-company-culture.html).

Charmaz, Kathy. 2008. "Grounded Theory as an Emergent Method" Pp. 155–172 in *Handbook of Emergent Methods* edited by Sharlene Nagy Hesse-Biber and Patricia Leavy. New York: Guilford Press.

Hesse-Biber, Sharlene Nagy and Patricia Leavy (editors). 2008. *Handbook of Emergent Methods*. New York: Guilford Press.

Introna, Lucas. 2017. "Phenomenological Approaches to Ethics and Information Technology." *Stanford Encyclopedia of Philosophy*June 29. Retrieved July 25, 2018 (https://plato.stanford.edu/entries/ethics-it-phenomenology/).

Kim, Jeong-Hee. 2016. "Narrative Data Analysis and Interpretation: Flirting with Data" Pp. 185–224 in *Understanding Narrative Inquiry*. Los Angeles: Sage.

Lester, Stan. 1999. "An Introduction to Phenomenological Research." Taunton UK: Stan Lester Developments. Accessed June 18, 2018 (https://www.researchgate.net/profile/Stan_Lester/publication/255647619_An_introduction_to_phenomenological_research/links/545a05e30cf2cf5164840df6.pdf).

MacNamara, Jim. 2005. "Media Content Analysis: Its Uses, Benefits and Best Practice Methodology." *Asia Pacific Public Relations Journal* 6(1):1–34.

Marvasti, Armir B. 2011. "What is Qualitative Research?" Pp. 1–15 in *Qualitative Research in Sociology*. London: Sage.

Roller, Margaret R. 2015. "Social Constructionism and Quality in Qualitative Research Design." *Research Design Review*February 11. Retrieved September 21, 2018 (https://researchdesignreview.com/2015/02/11/social-constructionism-quality-in-qualitative-research-design/).

Skalski, Paul D., Kimberly A. Neuendorf, and Julie A. Cajigas. 2017. "Content Analysis in the Interactive Media Age" in Pp. 201–242 in *The Content Analysis Guidebook*, 2nd Edition edited by Kimberly A. Neuendorf. Thousand Oaks, CA: Sage.

Wall, Sarah. 2008. "Easier Said than Done: Autoethnography." *International Journal of Qualitative Methods* 7(1):38–53.

INDEX

For Product Safety Concerns and Information please contact our EU representative GPSR@taylorandfrancis.com
Taylor & Francis Verlag GmbH, Kaufingerstraße 24, 80331 München, Germany

www.ingramcontent.com/pod-product-compliance
Lightning Source LLC
LaVergne TN
LVHW011122300926
849160LV00008B/476
9780367235109